Soil Microbiology and Biochemistry

E. A. Paul

Department of Crop and Soil Sciences
Michigan State University
East Lansing, Michigan

F. E. Clark

Agricultural Research Service
United States Department of Agriculture
and
Colorado State University
Fort Collins, Colorado

Soil Microbiology and Biochemistry

Academic Press, Inc.
Harcourt Brace Jovanovich, Publishers
San Diego   New York   Berkeley   Boston
London   Sydney   Tokyo   Toronto

ACADEMIC PRESS, INC.
San Diego, California 92101

United Kingdom Edition published by
ACADEMIC PRESS LIMITED
24-28 Oval Road, London NW1 7DX

Library of Congress Cataloging-in-Publication Data

Paul, Eldor Alvin.
  Soil microbiology and biochemistry.

  Includes bibliographies and index.
  1. Soil microbiology. 2. Soil biochemistry.
I. Clark, F. E. (Francis Eugene), Date.
II. Title.
QR111.P335   1988   631.4'6   88-7681
ISBN   0-12-546805-9   (alk. paper)

PRINTED IN THE UNITED STATES OF AMERICA
88 89 90 91   9 8 7 6 5 4 3 2 1

Contents

9. Reduction and Transport of Nitrate

10. Return of Nitrogen to Soil: Biological Nitrogen Fixation

11. Mycorrhizal Relationships

12. Phosphorus Transformations in Soil

13. Sulfur Transformations in Soil

14. Microbial Transformation of Metals

Preface

The driving force in the formation of the allied fields of soil microbiology and soil biochemistry was the need to know about the organisms and reactions occurring in soil. Among the natural processes initially of concern and that now have been given roughly a full century of investigation are those of carbon mineralization, nitrogen fixation, sulfur oxidation, mycorrhizal formation, nitrification, and denitrification. As knowledge has accumulated about these and other soil processes, so too has the need intensified for an integrated level of approach to applied problems in agronomy, forestry, and environmental science. That has been our objective in the current text. In the writing of the several chapters, a process-oriented approach is used to provide the student with a unified perspective on nutrient cycling and the fundamental soil processes that are driven by microorganisms.

The text is aimed at the advanced undergraduate already holding some background in biology and chemistry and with some knowledge of the environment, especially of soils. In a book such as this, it is impossible to present all the necessary background information. Chapter 2, on the soil habitat, and Chapter 4, on soil biology, are offered as summarizing orientations. Chapters 1, 3, and 5 also present background material. Supplemental reading lists and a brief glossary are also offered as a convenience to the student.

Problems posed by current stresses on global ecosystems emphasize the need for good management of agricultural and forestry resources. Advances being made in the genetic control of microbial capabilities offer great potential for altering interactions between host plants and microorganisms. Our hope for this text is that it will give the student an adequate introduction to soil microbiology and soil biochemistry, and especially an understanding of the basic microbial processes in soil. Given that, the student should be able to envision possibilities for field applications of forthcoming advances in our area of science.

Acknowledgments

The production of a volume such as this can only be accomplished with much assistance. We thank Scott Smith, William Horwath, Ken Horton, and David Harris for assistance in editing. The original artwork is credited to the craftsmanship of Phyllis Paul and Marlene Cameron. Linda Salemka is thanked for the manuscript preparation and placement onto the computer disk. Francis Clark, formerly an employee of the Agricultural Research Service, and currently a collaborator, thanks his co-workers for their many helpful suggestions.

ELDOR A. PAUL
FRANCIS E. CLARK

Chapter 1

Soil Microbiology and Biochemistry in Perspective

Definition and Scope

Soil microbiology is the study of organisms that live in the soil. The main thrust is on their metabolic activities and their roles in the energy flow and cycling of nutrients associated with primary productivity. Additionally, the discipline is concerned with the environmental impacts, both favorable and unfavorable, of soil organisms and the processes they mediate. To a large extent soil microbiology, if viewed from a mechanistic viewpoint, is soil biochemistry. The codisciplines were initially concerned with the microscopic life in the soil but became extended to include organisms of macroscopic size that reside in soil and participate in soil dynamics. Currently included in the soil biota together with the unicellular organisms are the soil-dwelling small invertebrates called the soil mesofauna. These may be either microscopic or macroscopic. Some protozoa are macroscopic, and many algae and fungi form communal or filamentous structures that are measurable in centimeters or decimeters.

The micro- and mesofauna play ancillary roles in organic matter transformations, but they lack the wide range of enzyme capabilities of the soil microflora. Inclusion of the small invertebrates within the scope of soil microbiology was only slowly accomplished; a similar delay occurred with respect to mycorrhizas. For many years only relatively few mycologists engaged in the study of the fungus–root association known as a mycorrhiza. As soil microbiologists became more concerned with nutrient transfers, they realized that the study of mycorrhizas was a central rather than a peripheral part of their discipline.

Pioneering Contributions

Although such phenomena as the spontaneous fermentation of fruit juices to yield wine and the souring of drawn milk have been observed by humans since the beginning of historical time, soil microbiology as a science can be assigned the same year of origin commonly given to bacteriology and protozoology. In 1676, the Dutch lens grinder Antonius van Leeuwenhoek reported that he had seen small animalcules in natural waters and in water in which pepper had lain. Inasmuch as his observations were on micro-organisms in the presence of decaying plant material, he could possibly be designated the father of soil microbiology. This designation, however, is commonly and justifiably given to Sergei Winogradsky (1856–1953) in recognition of his many contributions to the newly emerging science. Especially noteworthy was his discovery of the nitrifying bacteria and their role in the phenomenon of nitrification. This resulted in the concept of microbial autotrophy, wherein inorganic substrates are used as a source of growth energy by microorganisms.

Contemporaries of Winogradsky must be credited with another landmark discovery, namely, the formation of mycorrhizas by fungi and plant roots. Although earlier workers had noted the occurrence of fungus–plant root associations, it remained for Pfeffer (1877) to recognize the symbiotic nature of the association. A. B. Frank (1885) coined the term mycorrhiza; he later distinguished between ectotrophic and endotrophic mycorrhizas. The last half of the nineteenth century saw other important discoveries concerning microbial processes. These included symbiotic nitrogen fixation, denitrification, sulfate reduction, and asymbiotic nitrogen fixation.

The researches of Louis Pasteur (1830–1900) on microbial fermentations were of special significance; they led to the delineation of anaerobic metabolism. All multicellular forms of life, plant and animal, are dependent on an aerobic metabolism. Some soil bacteria are capable only of aerobic metabolism; others, only of anaerobic metabolism; while still others can switch from one form to the other. Several of Pasteur's predecessors had recognized that yeasts are involved in fermentations, but it remained for Pasteur to demonstrate that the production of alcohols and organic acids by microorganisms is linked to a basic metabolism that permits life without air. Büchner (1897) showed that yeast cells could be disrupted to yield a cell-free liquid capable of causing alcoholic fermentation; thus he must be credited for pioneer work in microbial enzymology.

In the years between the observations of Leeuwenhoek and those of Winogradsky, Pasteur, and contemporaries, there were general or background contributions, such as the disproving of spontaneous generation, the linking of microorganisms to plant and animal diseases, and great progress in the descriptive taxonomy of soil organisms. Linnaeus (1707–1778), the founder of binomial taxonomy, recognized the existence of microscopic forms of life but skirted a taxonomic quagmire by simply placing all microbes in a group designated "Chaos." Currently, microbial taxonomy remains controversial, but at least it has progressed from chaotic to utilitarian.

Soil Microbiology in the Early Twentieth Century

By the opening of the current century, the young science of soil microbiology was firmly established. Major research emphasis was on symbiotic nitrogen fixation, organic matter decomposition, and mineral nitrogen transformations. Successes in legume inoculations led to several premature attempts of other practical applications. Kluyver (1956) observed that "since Pasteur's startling discoveries of the important role played by microbes in human affairs, microbiology as a science has always suffered from its eminent practical implications." Around the turn of the century, considerable effort was expended in making census counts of soil organisms and attempting to use such counts as indices of soil fertility. This concept failed to take into account that, at best, the number of propagules capable of forming viable colonies on agar plates represents only a small percentage of the total microbial population, and also, that there are many other determinants of soil fertility, any one of which, under Liebig's law of the minimum, may restrict plant productivity. Attempts were also made to increase asymbiotic nitrogen fixation by inoculating nitrogenase-producing organisms into soil. These attempts failed because of the lack of knowledge concerning microbial competition. They, however, were at least partly responsible for transferring attention from the test tube to what microbes do or do not accomplish in the field.

The establishment of the general relationship between microbial growth and the transfers and transformations of organic nitrogen was among the early achievements. The carbon:nitrogen ratio required for plant residue degradation without a net tie-up (immobilization) of nitrogen was determined as approximately 25:1, and the effects of environmental factors on differential rates of plant decomposition were defined. The possibility of soil biota turnover with a subsequent release of nitrogen during decomposition was recognized. In summarizing the early work, Harmsen and van Schreven (1955) wrote, "The study of the general course of mineralization of organic nitrogen in soil was practically completed before 1935. It is surprising that many of the modern publications still consider it worthwhile to consider parenthetically observations dealing with those entirely solved problems." These authors, however, then pointed out that the relationships between carbon and nitrogen and the effects of environmental factors had to be determined for each soil type. This indicated that the underlying principles were not understood.

Progress and Diversification in Recent Years

With time, the scope of soil microbiology was gradually expanded from primary concern with nitrogen and organic matter to such areas as soil enzymes, the rhizosphere microflora, microbial participation in soil structure formation, degradation of manmade pesticides and other recalcitrants, microbial ecology, transformations of metals, and microbial impacts on the environment. Microbiologists participated in the empirical approach to agricultural productivity and range and forest management that coincided with the advent of cheap and available energy for tillage and fertilizer production during the period 1950–1980. This period also saw the development of tracer techniques. Norman (1946) wrote, "The availability of the stable nitrogen isotope ^{15}N and the carbon isotope ^{13}C will make it possible to verify quantitatively the various nitrogen transformations in relation to the carbon cycle and should aid greatly in establishing the forms of nitrogen present in the soil."

Jansson (1958) noted the preferential utilization of NH_4^+ rather than NO_3^- by microorganisms and the feasibility of using mathematical equations to describe mineralization–immobilization interactions. The literature in the 1960s documented the introductory work on the characterization of soil organic matter (SOM) into biologically meaningful fractions in addition to the classical fractionation techniques that provided humic and fulvic acids. The concept of a small, nutritionally active fraction with a significantly faster turnover than that of the large recalcitrant fraction greatly improved the interpretation of SOM dynamics and led to realistic mathematical models of SOM turnover. It also assisted in the development of soil tests for the potential bioavailability of nitrogen. The recognition of large and significant organic phosphorus and sulfur cycles coincided with the incorporation of soil biology and enzymology into ecosystems research.

Significant contributions to genetics, including bacterial genetics, occurred during the first half of this century. Griffith (1928), presented the first evidence of bacterial transformation by showing that avirulent, noncapsulated cells of *Streptococcus pneumoniae* could be transformed to virulent, capsulated cells by injecting host animals with heat-killed virulent cells together with live, avirulent cells. In 1941, N. A. Krassilnikov published results showing the transformation of a noninfective *Rhizobium* to an infective one. There was, however, some question about the purity of his inoculum. The work of Avery *et al.* (1944) on the chemical nature of the substance inducing transformation was followed by the Nobel Prize–winning description of the structure of deoxyribonucleic acid (DNA) by Crick and Watson in 1953. The mechanism of heredity control was shown to be a sequence of four nucleotides arranged as a two-stranded molecule in a double helix. The work in the mid-1970s on plasmid biology, DNA sequencing (the determination of the specific sequence of nucleotide bases), and the discovery of transcription enzymes set the stage for genetic engineering techniques. These have great potential for use in agricultural and environmental microbiology. Some of the possibilities are discussed in later paragraphs.

Areas of interest in soil microbiology and shifts therein within soil science are reflected in the Division III publications in the *Soil Science Society of America Journal* during the years 1946–1985 (Fig. 1.1). The histograms show only six categories, but "miscellaneous" covers a dozen or so topics (e.g., fauna, rhizosphere, enzymes, census counts, antibiotics, mycorrhizas), any of which, if classed separately, would include only about 1% of the total number of publications in each decade. Over the time span covered, studies on nitrogen fixation were at a low ebb for the middle decades, during which time studies on soil structure and on pesticide degradations peaked and then subsided. Emphasis on soil structure studies occurred during years in which much attention was being given to the use of soil additives for promoting soil structure formation. Emphasis on pesticide studies occurred during years characterized by environmental concerns and, in the United States, the establishment of the Environmental

Figure 1.1. Soil microbiology publications in the *Soil Science Society of America Journal* for the years 1946–1985.

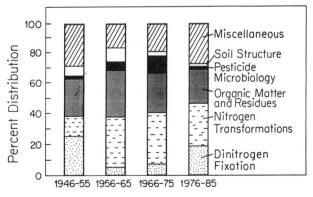

Protection Agency and the enactment of clean air and clean water legislation. Figure 1.1 shows that publications in soil nitrogen transformations comprise the largest single subgroup in three of the four histograms. This stems from the popularity of nitrogen-15 studies after 1955 and of denitrification studies, especially N_2O emissions, after 1975. The shifts in emphasis over time attest that soil microbiology is an applied science.

Literature of Soil Biology

Prior to 1900, reports on research in soil microbiology were published in diverse journals having no titular reference to soil microbiology as such. The first instance of a major journal giving specific subtitle reference to the discipline occurred in 1895. *Centralblatt für Bakteriologie und Parasitenkunde Infektionskrankheiten und Hygiene* entitled its Abteilung 2 as "Allegemeine, landwirtschaftliche und technische Bakteriologie, Garungsphysiologie und Pflanzenpathologie." Listing all the journals in which soil microbiological and biochemical research is currently reported would require hundreds of entries. Two publications making sectional provision for soil microbiology at their initiation were the *Soil Science Society of America Proceedings* (Volume 1 in 1937) and *Transactions of the International Society of Soil Science* (first congress in 1927). The microbiological subsections were entitled "Soil Microbiology" and "Soil Biology," respectively.

An interesting development in the late 1960s was the formal wedding of soil microbiology and biochemistry. In 1968 the Soil Science Society of America changed its Division III designation from "Soil Microbiology" to "Soil Microbiology and Biochemistry." Almost simultaneously (1969), there appeared a new international journal entitled *Soil Biology and Biochemistry*. Starting in 1967, there was published a five-volume series entitled "Soil Biochemistry"; these volumes (1967–1981) were written primarily for a readership of soil microbiologists.

Linkages of soil microbiology with other soil disciplines continue. In 1985 there appeared the first volume of a journal entitled *Biology and Fertility of Soils*. Symposia, workshops, and conferences involving soil biology with diverse other disciplines now commonly occur.

The first major textbook for soil microbiology was written in German. In 1910 Felix Löhnis published his "Handbuch der Landwirtschaftliche Bakteriologie." This text reached a total of 19 editions and was translated into four other languages. Other noteworthy early texts were E. J. Russell's "The Microorganisms of the Soil" in 1923 and Waksman's "Principles of Soil Microbiology" in 1927. A less cumbersome volume entitled "The

Soil and the Microbe'' by Waksman and Starkey was published in 1931. Since then, a number of texts written in English have been published (Alexander, 1961, 1977; Gray and Williams, 1971; Hattori, 1973; Walker, 1975; Hawker and Linton, 1979; Subba Rao, 1982; Lynch, 1983; Aiken *et al.*, 1985).

Looking toward the Twenty-first Century

It is now approximately 100 years since the isolation by Beijerinck in 1888 of the symbiotic nitrogen-fixing bacteria now known as *Rhizobium*. The establishment of the principles of nitrification and the autotrophic way of life by Winogradsky (1890) could probably be said to be the start of soil biochemistry. A number of new factors are influencing soil microbiology and biochemistry as it enters its second century. These include (1) the use of computers for data processing, information exchange, and the control of automated instrumentation, (2) the application of whole-system management techniques, such as zero-till, alternate cropping, and whole-tree harvesting, (3) the influence of genetic engineering, (4) the realization that soil microbiological processes are affected by and in turn influence a number of major environmental problems, and (5) the need to develop more highly efficient agriculture and forestry management systems.

Genetic engineering techniques can be used to enhance nitrogen fixation, increase microbial capabilities for decomposing recalcitrant materials, and achieve new biological controls for agricultural pests. As newly engineered organisms are introduced into soil, there is also the possibility of environmental catastrophy. For better or worse, soil microbiology and biochemistry are closely involved with biotechnology. The word biotechnology is variously defined—simply, as the study of cellular and molecular biology, and more verbosely, as the integrated use of biochemistry, microbiology, and engineering sciences in order to achieve the technological application of the capacities of microorganisms, cultural tissue, and cells and parts thereof. Lynch (1983) defined soil biotechnology as the study and manipulation of soil organisms and their metabolic processes to optimize crop productivity. His definition could well be amended to include manipulations to optimize the quality of the environment. Microbiota are increasingly called on to degrade new organic molecules and to rid the soil and surface water of pollutants ranging from oil spills to excess nitrates to sewage constituents.

In soil microbiology's second century, one of the greatest challenges will be to produce and to manage newly engineered microorganisms. To be of use in nature, an engineered organism must be amenable to successful inoculation into the natural environment and to survive in competition with the broad range of native organisms. To offset the very real fears that any new organism could become a pest, it should not be able to escape to environments other than that for which it was produced. It should also not have the potential for genetic interchange with native soil organisms to produce unknown organisms. The potential for transfer of new genetic material has caused extensive discussion and slowed the testing of newly structured organisms in the field. The fear of an undesired spreading of new organisms stems partly from experience with previous chance introductions, such as the gypsy moth and Dutch elm disease into America. In the past, large amounts of inoculants of naturally occurring organisms have been applied to soil in order to enhance desirable symbioses or to foster biological controls. To date, there have been no known adverse effects of such inoculations. A review of the early literature on the fate and survival of genetically engineered organisms has been given by Halvorsen *et al.* (1985).

The utilization of genetically engineered organisms in environmental improvement involves a knowledge of the basic characteristics of soil organisms. Enhanced degradation of resistant toxic organics and the rapid destruction of pathogenic contaminants would have major cost–benefit effects in sewage treatment. Agricultural pesticides are reaching the ground water in many areas of the world. Detoxification of these ground waters is difficult because of low carbon supplies for microbial growth and often adverse abiotic conditions. The high pollution potential of pesticides makes this an important field for research. Chlorinated hydrocarbons, which form a significant portion of pesticides, have generally been found to be poorly degraded under anaerobic conditions such as those existing in sediments, rice fields, and lower soil strata. The development of bacteria to degrade chlorinated hydrocarbons under anaerobic conditions, therefore, has potential for biotechnological breakthrough. The demonstration that white-rot fungi, which are known to degrade lignins, also can metabolize the toxic, very resistant, organic polychlorinated biphenyls (PCBs) is a further example of the diversity of soil microflora. It also is an example of the application of biotechnology in its broader sense to environmental problems.

The high application rates of pesticides to much of our agricultural land could be lowered and, it is to be hoped, in some cases eliminated through better biological control mechanisms. These include a variety of techniques, such as integrated pest management, altered tillage practices, and the development of more pest-resistant plants. Genetically altered microorganisms utilized as biological control agents and newer methods of breeding plants would be part of a management scenario based on a better understanding of soil microbiology and biochemistry.

References

Aiken, G. A., McKnight, D. L., Wershaw, R. L., and MacCarthy, P. (1985). "Humic Substances in Soil, Sediment and Water: Geochemistry, Isolation and Characterization." Wiley, New York.

Alexander, M. (1961). "Introduction to Soil Microbiology." Wiley, New York.

Alexander, M. (1977). "Introduction to Soil Microbiology." 2nd ed. Wiley, New York.

Avery, O. T., MacLeod, C. M., and McCarty, M. (1944). Studies on the chemical nature of the substance inducing transformation of pneumococcal types. *J. Exptl. Med.* **79**, 137–158.

Beijerinck, M. W. (1888). Die Bakterien der Papillonaceen-Knollchen. *Bot. Ztg.* **46**, 724–735.

Büchner, E. (1897). Alkaholische Gährung ohne Hefezellen. *Ber. dtsch. chem. Ges.* **30**, 112–124.

Crick, F. H. C., and Watson, J. D. (1953). Molecular structure of nucleic acid. A structure for deoxyribose nucleic acid. *Nature, Lond.* **171**, 737–738.

Frank, A. B. (1885). Ueber die auf Wurzelsymbiose beruhende Ernährung gewisser Bäume durch unterirdische Pilze. *Ber. deut. bot. Ges.* **3**, 128–145.

Gray, T. R. G., and Williams, S. T. (1971). "Soil Microorganisms." Oliver & Boyd, Edinburgh.

Griffith, F. (1928). The significance of pneumococcal types. *J. Hygiene* **27**, 113–159.

Halvorsen, H. O., Pramer, D., and Rogul, M. (1985). "Engineered Organisms in the Environment: Scientific Issues." Am. Soc. Microbiol., Washington, D.C.

Harmsen, G. W., and van Schreven, D. A. (1955). Mineralization of organic nitrogen in soil. *Adv. Agron.* **7**, 299.

Hattori, T. (1973). "Microbial Life in the Soil." Dekker, New York.

Hawker, L. E., and Linton, A. H. (1979). "Microorganisms, Function, Form and Environment." Edward Arnold, London.

Jansson, S. L. (1958). Tracer studies on nitrogen transformations in soil with special attention to mineralization-immobilization relationships. *Kungl. Lantbrukshogskolan Ann.* **24**, 101–361.

Kluyver, A. J. (1956). The microbe's contribution to biology. Harvard University lecture. Cambridge.

Krasil'nikov, N. A. (1941). Variation in nodule bacteria. *Mikrobiologiya* **10**, 396–400.

Löhnis, F. (1910). "Handbuch der Landwirtschtliche Bakteriologie" Borntraeger, Berlin.

Lynch, J. M. (1983). "Soil Biotechnology." Blackwell, Oxford.

Norman, A. G. (1946). Recent advances in soil microbiology. *Soil Sci. Soc. Am. Proc.* **11**, 4–15.

Pfeffer, W. (1877). Ueber fleischfressende Pflanzen. *Landw. Jahrb.* **6**, 969–998.

Russell, E. J. (1923). "The Microorganisms of the Soil." Longmans, Green. London and New York.

Subba Rao, N. S., ed. (1982). "Advances in Agricultural Microbiology." Butterworth, London.

Walker, N. (1975). "Soil Microbiology: A Critical Review." Butterworth, London.

Waksman, S. A. (1927). "Principles of Soil Microbiology." Williams & Wilkins, Baltimore, Maryland.
Waksman, S. A., and Starkey, R. L. (1931). "The Soil and the Microbe." Wiley, New York.
Winogradsky, S. 1890. Recherches sur les organisms de la nitrification. *Ann. Inst. Pasteur* **4**, 213–31, 257–75, 760–71.

Supplemental Reading

Brill, W. J. (1985). Safety concerns and genetic engineering in agriculture. *Science* **227**, 381–384.
Dykhuizen, D., and Hartl, D. (1981). Evolution of competitive ability in *Escherichia coli. Evolution (Lawrence, Kans.)* **35**, 581–594.
Kolata, G. (1985). How safe are engineered organisms? *Science* **229**, 34–35.
Kosuge, T., and Nester, E. W. (1984). "Plant Microbe Interactions: Molecular and Genetic Perspectives," Vol. 1. Macmillan, New York.
Lincoln, D. R., Fisher, E. S., and Lambert, D. (1985). Release and containment of microorganisms for applied genetic activities. *Eng. Microb. Technol.* **7**, 314.
Paul, E. A., and Ladd, J. N., eds. (1981). "Soil Biochemistry," Vol. 5. Dekker, New York.
Schnitzer, M., and Khan, S. U. (1972). "Humic Substances in the Soil Environment." Dekker, New York.
Tangley, L. (1985). Releasing engineered organisms in the environment. *BioScience* **35**, 470–473.

Chapter 2

Soil as a Habitat for Organisms and Their Reactions

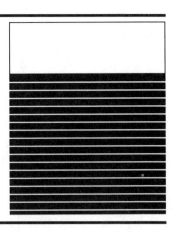

Introduction

Soil organisms participate in the genesis of the habitat wherein they live. They, together with the total biota and especially the higher vegetation, constitute one of the five interactive factors in soil formation; the other four are climate, topography, parent material, and time. The physical and chemical breakdown of rocks to fine particles with large surface areas and the accompanying release of plant nutrients initiate the soil-forming process (Fig. 2.1). Two major nutrients that are deficient in the early stages of the process are carbon and nitrogen; therefore, the initial colonizers of soil parent material are usually organisms capable both of photosynthesis and nitrogen fixation. These are predominantly the cyanobacteria, also known as the bluegreen algae. After higher vegetation has become established, a continuum of soil processes produces the dynamic mixture of living and dead cells, soil organic matter (SOM), and mineral particles in sufficiently small sizes to permit the intimate colloidal interactions characteristic of soil.

The interaction of vegetation–time factors is most easily studied in new landscapes. The eruption of the volcano Krakatoa in 1883 provided small, neighboring Indonesian islands with a new coating of parent material. The progress of revegetation and the process of soil genesis are still continuing on these islands. The volcanic island, Surtsey, formed off Iceland in the 1960s, and the posteruptive landscape around Mt. St. Helens in the 1980s provide more current examples of the initiation of soil formation. The revegetation of coal strip-mine spoils and of newly formed littoral sand dunes also exemplify areas basically devoid of life becoming stabilized by the development of soil through vegetation and microbial activity.

Figure 2.1.   Interrelationships of organisms, organic matter, and parent materials in soil development.

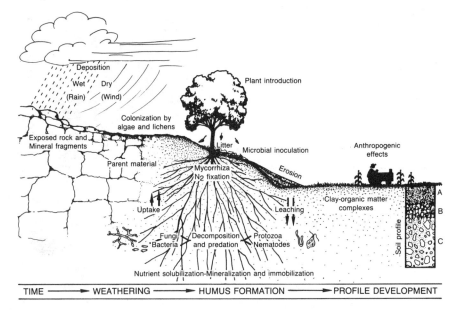

A modern definition of soil relates to the earth's surface layer exploited by plant roots. Microbial growth occurs to much greater depth. Living organisms are found in oil wells, amidst rocks where they have been protected from surface environments for millions of years. Active denitrification occurs in subsoils much below the rooting depth if a source of carbon is percolated downward with the NO_3^-. There appear to be no known natural areas on the earth's surface, with the exception of active volcanoes, where microbial life is absent. Microbes are present in the Gobi desert, where diurnal temperature fluctuations can attain 50°C, at the polar ice caps at −50°C, and in hot springs at 90°C. Viable bacteria also have been said to be found in the Galapagos Trench, where ocean water is under great pressure and high temperature.

The soil microbiologist is primarily concerned with those areas of the earth's surface that when undisturbed, constitute major terrestrial ecosystems, and when managed, provide arable soils or other resources. Soil organisms show their greatest diversity of species and usually their largest populations in productive soils. The size of the microbial biomass usually shows direct correlation with the amount of plant growth (the primary productivity) and with SOM levels. Good discussions of the microbial environment have been given by Chen and Avnimelech (1986), Lynch and Poole (1979), McLaren and Skujins (1968), and Focht and Martin (1979).

Structural Aspects of Soil

Soil consists of mineral particles of various sizes, shapes, and chemical characteristics, together with plant roots, the living soil population, and an organic matter component in various stages of decomposition. Soil gases, soil water, and dissolved minerals complete the soil habitat. An understanding of the soil fabric requires a knowledge of the spatial arrangements. These relate to both the size and shape of the components. The relative dimensions of components of the soil matrix (Fig. 2.2) range from 2 mm or greater for macroaggregates to fractions of a micrometer for bacteria and colloidal particles. Enzymatic and other molecular reactions occur at size dimensions at least another order of magnitude smaller.

The formation of clay–organic matter complexes and the stabilization of clay, sand, and silt particles into aggregates are the dominant structural features of most soils. Clays are basic to aggregate formation. The pedologist describes clay particles as being less than 2 μm in diameter and refers to them as colloidal. In chemistry, colloids are characterized by their dispersion and large surface area. Although a large percentage of clay particles are larger than many other colloidal materials as the chemist knows them, because of their large surface area, they act as colloids in nature. Clays such as smectite have a surface area of approximately 400 $m^2 g^{-1}$ per unit interlayer of the clay.

Most clays have a net negative charge. Microorganisms also are negatively charged at neutral pH values, as are most SOM constituents. Normally, two negatively charged units would repel each other. Attachments between negatively charged units are possible by ionic bonding via multivalent cations. One of the bonds attaches to a microorganism or organic matter and another to clay. Microbial polysaccharides and fibrils with strong attachment characteristics also bind soil particles together. Many tropical soils, especially those derived from volcanic rocks, have allophanic clays of variable charges. These clays contain high concentrations of iron and aluminum that help to stabilize SOM constituents. Many of such soils have an excellent soil structure and a SOM content two or three times as great as would be encountered in soils developed under similar climatic conditions but from nonallophanic parent materials.

Soil aggregation is one of the most important factors controlling microbial activity and SOM turnover. Aggregate formation is initiated when microflora and roots produce filaments and polysaccharides that combine with clays to form organic matter–mineral complexes. Soil structure is created when physical forces (drying, shrink–swell, freeze–thaw, root growth, animal movement, and compaction) mold the soil into aggregates. Figure 2.3 shows microorganisms within an aggregate. Although this

Figure 2.2. Model of aggregate organization, showing relative size of the components and the major binding agents. (Adapted from Tisdall and Oades, 1982.)

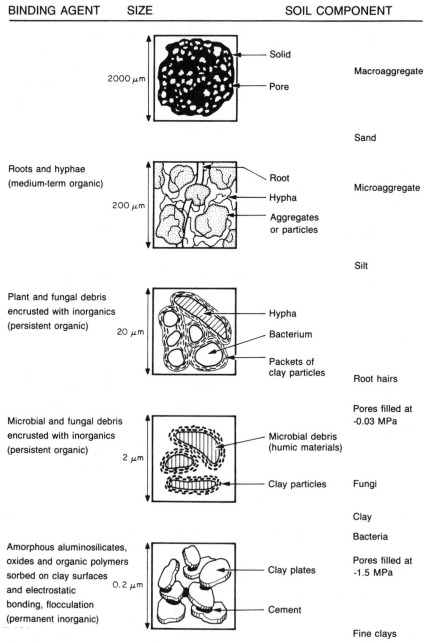

Figure 2.3. Model of soil aggregate, showing organic matter protected from attack by crypt formation.

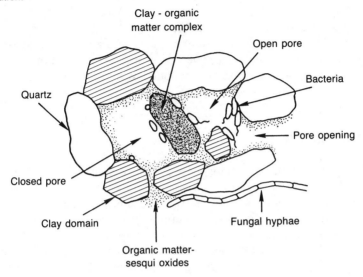

Clay - organic matter complex

Open pore

Bacteria

Quartz

Pore opening

Closed pore

Clay domain

Fungal hyphae

Organic matter-sesqui oxides

diagram is simplified and size relationships somewhat altered, it shows how pore sizes are affected by distances between particles and how organic matter may be protected. Organic matter in portions of the aggregate will not be mineralized if it is physically separated from the microorganisms and their enzymes.

Most organisms exist on the outside of aggregates and in the small pore spaces between them; relatively few reside within the aggregate. Those that do usually remain there from the time of its formation to the time of its disruption. Pictures from electron microscopy and calculations of microbial density indicate that microorganisms occupy less than 1% of the total available pore space. Pore neck sizes determine accessibility to pores by organisms according to their body sizes. Occupancy is also affected by the water content of the pore. Pores as small as a very few micrometers in diameter and filled with water are generally suitable for bacteria. Fungi invade somewhat larger pores. Pore size limits the ability of the fauna to move about; in turn, this affects their grazing on the soil microflora. Microaggregates (50–250 μm) are subunits of macroaggregates (>2 mm), the interstices of which form macropores. The macropores provide refuges for the microfauna, protecting them from larger predatory organisms.

Soil aggregates and their constituent clays influence the interaction of enzymes with their substrates. The clay particle with its large external and internal surface areas is capable of adsorbing enzymes such as urease and protease. Enzymes adsorbed to clay or intertwined with humate constituents are protected from hydrolysis by other enzymes. Adsorption also makes the catalytic site less available. A small molecule, such as urea (NH_2CONH_2), can diffuse readily to a urease site and there undergo decomposition. A large molecule, such as a protein, would not diffuse as readily to a protease site and consequently would be broken down at a much slower rate than would urea.

Soil Atmosphere

The major gases in the soil atmosphere are those found in the atmosphere, namely, N_2, O_2, and CO_2. Gases arising from biological activity, such as nitrogen oxides, may at times be present; because of their high reactivity with soil components and their susceptibility to biological activity they are usually transitory. In well-aerated soils, the O_2 content seldom falls below 18 to 20%, and CO_2 seldom rises above 1 to 2%. However, given a clay texture and high moisture content coupled with high microbial activity, CO_2 content of the soil atmosphere may reach as high as 10%.

The diffusion of gases in soil is described by Fick's Law which relates gas movement to soil characteristics, the concentration of the gas in the soil and the depth as follows:

$$q_i = D_{sa,i} \frac{dc_i}{dz}$$

where q_i is the diffusion rate (g cm^{-2} sec^{-1}) of gas, $D_{sa,i}$ the diffusion constant (cm^2 sec^{-1}) in soil, c_i the concentration (g cm^{-3}) in soil air, and z the depth (cm).

The solubility of gases in water depends on the type of gas, temperature, salt concentration, and the partial pressure of the gases in the atmosphere. The most soluble gases are those that become ionized in water, e.g., CO_2, NH_3, and H_2S. Oxygen is markedly less soluble (Table 2.1), and N_2 even less so. Nitrogen can become limiting in nitrogen-fixing systems because of its low aqueous diffusion rate even though 80% of the atmosphere is N_2. The low diffusion rate of N_2 into poorly ventilated sites also affects the measurement of N_2 fixation by the acetylene (C_2H_2) reduction technique; C_2H_2, being soluble in water, is much more available at depth than N_2. This can lead to erroneously high estimates of N_2 fixation when using C_2H_2 reduction as an indicator of the reaction rate. Some plants, such as rice and certain bog plants, possess special root channels that provide paths for the diffusion of air into the submerged layers of the soil. Rice

Table 2.1
Diffusion Constants of CO_2, O_2, and N_2, in Air and Water,
and Their Solubility in Water at 20°C

| | Diffusion constant ($cm^2 sec^{-1}$) | | Solubility in H_2O ($cm^3 liter^{-1}$) |
	Air	Water	
CO_2	0.161	0.177×10^{-4}	8.878
O_2	0.205	0.180×10^{-4}	0.031
N_2	0.205	0.164×10^{-4}	0.015

plants provide four times as great a diffusion of air into lower horizons than does barley. It thus produces aerobic microsites around its roots in an otherwise anaerobic zone when grown under flooded conditions.

The extent of aeration of soil can be inferred from the water content. Mineral particulates usually have a density of 2.65 g cm^{-3}. The bulk density of surface soil generally ranges from 0.9 to 1.3 g cm^{-3}. Therefore, mineral soils usually consist of 50 to 60% by volume of pores of assorted sizes. The water content of a soil at -0.01 megapascals (MPa) (field capacity) ranges from 15 to 30% for sandy loams to 40 to 45% for clays. The difference between the total pore space and the water content represents the air-filled space. A minimum air-filled pore space of 10% by volume is commonly considered necessary for adequate aeration. Clay soils with 45% water may not have enough air spaces for aeration if their bulk density is 1.3 and total pore space is only 50%.

It has been determined that the change from aerobic to anaerobic metabolism occurs at O_2 concentrations of less than 1%. The overall aeration of a soil is not as important as that of the individual crumbs and aggregates. Calculations show that water-saturated soil crumbs larger than 3 mm in radius have no O_2 in the center (Harris, 1981). The fact that anaerobic processes such as denitrification and sulfate reduction occur in many soils indicates that anaerobic microsites must occur quite commonly. Further evidence that anaerobic microsites exist can be deduced from the common occurrence of anaerobic bacteria such as clostridia in the upper layers of soil. Several studies have shown that the population of anaerobic bacteria in the upper few centimeters of soil can be as much as 10 times their number at greater depths. Aerobic bacteria play a preparatory role in producing an environment for the anaerobes. Their initial growth within a microsite consumes its stored O_2, thus allowing anaerobes to develop.

Anaerobic microorganisms have the ability to generate energy and grow in the absence of O_2. Some organisms (the obligate anaerobes) are also poisoned by the presence of O_2. This inhibition results from the production of toxic intermediates, as shown in Fig. 2.4. Organisms not sensitive to

Figure 2.4. Toxic intermediates O_2^- and H_2O_2, formed from single-electron reductions of O_2. The enzymes which protect aerobic organisms from these toxicants are also shown. Anaerobes often lack these protective enzymes. (From Tiedje et al., 1984.)

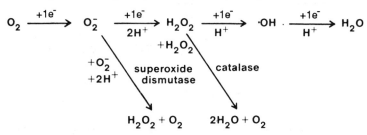

O_2 contain enzymes (superoxidases and/or catalase) to remove the toxic intermediates. Since the toxic intermediates are produced during electron transport, the inhibition is noted only in the presence of an electron donor (food supply). Obligate anaerobes can persist for long periods in an aerobic soil environment if they do not have adequate substrate for growth; such is often the case in soils.

Some soil organisms have become adapted over geologic time to the enhanced level of CO_2 that exists in the soil atmosphere. Certain species of fungi show preference for the 10- to 20-cm depth of soil. The nitrifying bacteria also show preference for CO_2 levels greater than that found above ground. With experimentally controlled atmospheres in closed containers, little nitrification occurs with CO_2 at the atmospheric level of 0.035% compared to that occurring at the 0.07 to 0.23% commonly present in soil.

Profiles of CO_2 contents in soils that do not have a water table near the surface show that the CO_2 concentration is highest near the surface, where root and microbial respiration occur. Root respiration contributes from 20 to 50% of the soil CO_2 flux (Table 2.2). The greater fraction of 50 to 80% can be attributed to microbial effects. In the absence of respiratory activity, as in frozen or dry soils, soil gases soon equilibrate with the atmosphere.

Table 2.2
Contribution of Roots to Soil Respiration

	Root respiration (% of total)
Oats	30
Wheat	20
Prairie grass	19
Oak trees	40

Excess water in soil restricts microorganisms and their activities by preventing O_2 movement into and through the soil in sufficient quantity to meet the O_2 demand of the organisms. With soil water contents such as those commonly found in cultivated field soils, O_2 moves into the soil largely by diffusion from the atmosphere. The gas pressure difference needed to move O_2 into the soil is only of the order of 1 to 3%. Thus O_2 content of agricultural soils rarely falls below the level of 18 to 20%. Similarly, once the level of respiratory CO_2 accumulation in the soil reaches 1 to 3%, a sufficiently large diffusivity gradient is established to move soil CO_2 into the atmosphere (0.035% CO_2). With water-sealing of the soil surface, such as may occur during the application of irrigation water or during downpours of rain, movement of atmospheric gases to and from soil almost completely stops. This is not surprising, inasmuch as O_2 diffuses almost 10,000 times as fast through air spaces as it does through water. Microbial activity does not become curtailed concurrently with water-sealing of the soil surface, but only after the reservoir of O_2 entrapped below the soil surface is exhausted. The size of this reservoir and the length of time it would be adequate for plant root and microbial respiration can be calculated from a knowledge of soil porosity, content of O_2 in the soil air, depth to the water table, and rate of O_2 use by plant roots and microorganisms.

Soil Water

Soil water affects not only the moisture available to organisms but also the soil aeration status, the nature and amount of soluble materials, the osmotic pressure, and the pH of the soil solution. The shape of the water molecule, with an HOH angle of 105°, results in the side with the hydrogen being electropositive whereas the other is electronegative. This explains many of water's properties relative to physical and chemical reactions. It explains why water is attracted to charged ions. Cations such as Na^+, K^+, and Ca^{2+} become hydrated because of their attraction to the negatively charged oxygen end of the water molecule. The polar nature of water also explains hydrogen bond formation by water. The bonding of each water molecule to other water molecules and to other biological components (Fig. 2.5) explains the solution properties, viscosity, and high specific heat of soil water. Of special significance within the microbial cell, as well as in the soil system, is the fact that water adsorbs strongly to surfaces by hydrogen bonding and dipole interactions. The thin layer of adsorbed water remains unfrozen at 0°C, and its removal by heating requires a temperature

Figure 2.5. Hydrogen bonds of importance to soil biological and organic matter structures.

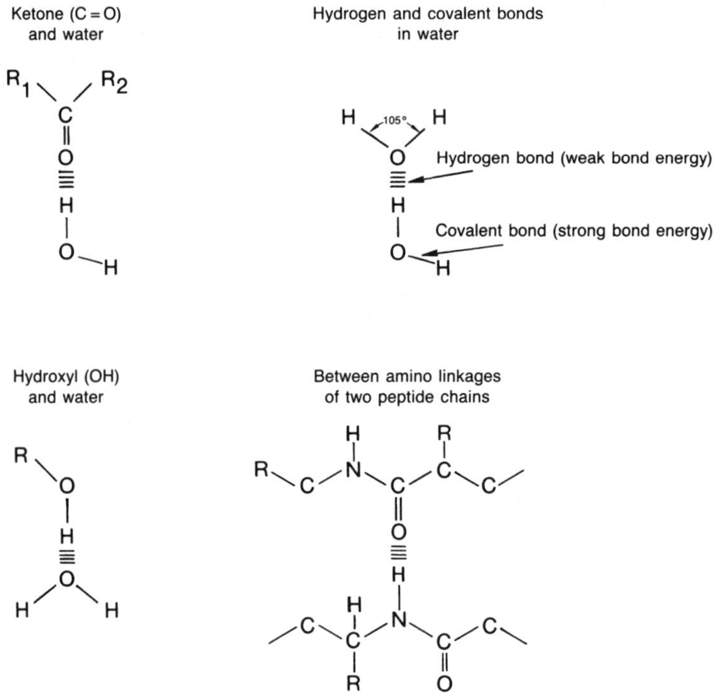

Ketone (C = O) and water

Hydrogen and covalent bonds in water

Hydrogen bond (weak bond energy)

Covalent bond (strong bond energy)

Hydroxyl (OH) and water

Between amino linkages of two peptide chains

up to 105°C. This bound water therefore has characteristics quite different from that of unbound water.

The presently accepted terminology for soil water characteristics is based on the concept of matric and osmotic potentials. Matric potentials are attributed to the above-described attraction of water to solid surfaces. Since this reduces the free energy of water, matric potentials are negative. Solutes in the soil, because of water's solution properties, also reduce the free energy of water and create another negative potential, the osmotic potential. The concept of soil water potential relates to basic definitions of the free energy involved and is expressed in pascals (usually given in megapascals).

Soil water has commonly been described as existing in three forms: gravitational, capillary, and hygroscopic. The first two terms are largely self-explanatory. Gravitational water is drawn through the soil by gravitational forces. This will occur after irrigation or a heavy rain. Immediately after gravitational water has drained away, the soil water is at field capacity. The micro-, or capillary, pores are, however, still filled with water that is available for plant and microbial growth. The matric potential of this water will be −0.01 to −0.03 MPa. As water is lost from soil either by evaporation or transpiration, plants will not be able to obtain enough to remain turgid both night and day and will be said to be wilted. This commonly occurs near −1.5 MPa and is said to be the wilting point. Hygroscopic soil water is variously defined as that absorbed by a dry soil from an atmosphere of high relative humidity, as that remaining in soil after air-drying, or as that held by a soil when in equilibrium with a specified relative humidity (RH) and temperature, usually 98% RH and 25°C. These three definitions are now obsolete, as is the term pF (logarithm of the soil water tension expressed in centimeter height of a column of water). The description of soil water availability relative to its matric potential is a much more useful concept because it can be reproducibly measured and is scientifically based.

The combined matric and osmotic components of soil water determine the stress against which an organism must work to obtain water. Generally, microbial activity in soil is optimal at −0.01 MPa and decreases as the soil becomes either waterlogged near zero water potential or more droughty at large, negative water potentials (Fig. 2.6). The curve showing the relative reaction rate as a percentage of maximum is for soil organisms collectively in reactions such as decomposition, with the highest rate being assigned a value of 100%. It does not apply uniformly to different groups of organisms, nor to individual species within a group. Fungi are generally

Figure 2.6. Relative microbial reaction rates at various moisture stresses. Because sands and clays have much different retention characteristics, the curve would occur at much lower water contents in a sandy soil than in a clay.

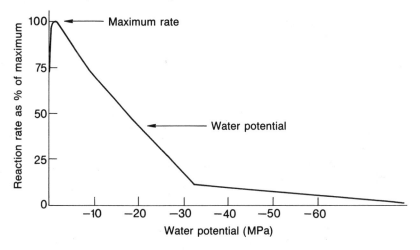

Table 2.3
Upper Tolerance Levels or Microorganisms to Water Potential Controlled by Solute
Concentration[a]

Designation of water potential			Solution concentration		
MPa	bar	A_w	NaCl (w/v)	Sucrose (w/v)	Organism
−1.5	−15	0.99	2.0	17	*Rhizobium, Nitrosomonas*
−10	−100	0.93	12.3	52	*Clostridium, Mucor*
−25	−250	0.83	25.3	70	*Micrococcus, Penicillium*
−65	−650	0.62	—	83	*Xeromyces, Saccharomyces*

[a]From Harris (1981).

more tolerant of higher water potentials (greater water stress) than are bacteria. Table 2.3 shows differences, among microorganisms, in their upper tolerance levels to moisture potential. The nitrifiers, as typified by *Nitrosomonas,* are less tolerant of stress than are the ammonifiers, as typified by *Clostridium* and *Penicillium.* Ammonia may accumulate in droughty soil because the nitrifiers cannot operate at water potentials at which the ammonifiers such as *Penicillium* are still active. The enhanced NO_3^- content sometimes observed in the surface layer of a droughty soil may not be due to nitrification but to upward movement of capillary water carrying NO_3^-. The evaporation of the water component at the soil surface leaves the NO_3^- behind.

Water influences both microorganisms and higher plants through the effects of diffusion, mass flow, and the concentration of nutrients. At high but not limiting moisture stress, nutrient diffusion may be sufficiently slow to be limiting. Mass flow usually is adequate to supply nutrients such as NO_3^- to plant roots. Phosphorus, however, does not move readily with the water; its uptake requires diffusion and root extension.

Redox Potential

Reduction–oxidation reactions are of major significance in explaining both soil chemical and biological phenomena and should receive much more detail than can be given in this overview of factors controlling soil biota. The following short explanation stresses the biological component. For further information, the suggested supplemental reading should be consulted. A comprehensive discussion of redox reactions has been given by Stumm and Morgan (1981).

Life obtains its energy from the oxidation of reduced materials, i.e., it removes electrons from either organic or inorganic substrates to capture the energy that is available during oxidation. This is accomplished in a series of steps involving a number of intermediate reactions. Electrons from reduced compounds are moved along respiratory or electron transport chains composed of a series of components. A chain capable of moving electrons from substrates to O_2 as the final electron acceptor is shown in Fig. 2.7.

In some cases where O_2 is not available, NO_3^-, Fe^{3+}, Mn^{2+}, and SO_4^{2-} can act as electron acceptors if the organism has the appropriate enzyme systems. Figure 2.7 shows that the orientation of the electron transport system is asymmetric, resulting in a separation of the movement of electrons and protons. The protons are expelled outside the membrane. The electrons are eventually transferred to O_2, forming OH^- on the inside. Since the membrane is not freely permeable to the H^+ on the outside and

Figure 2.7. Bacterial electron transport chain, showing the series of membrane-bound reactions that can result in the transfer of electrons to O_2, resulting in the generation of ATP. The proton gradient produced facilitates energy production and the production of H^+ outside the membrane.

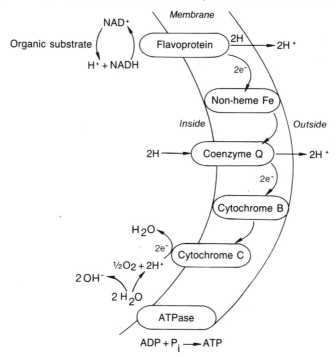

the OH^- inside, a pH gradient and electrical potential are established across the gradient. This proton motive force can do work such as active transport and flagella rotation, or it can be used to generate energy through ATPase to form ATP.

Just as acids and bases have been spoken of as proton donors or acceptors, reductants and oxidants are defined as electron donors or acceptors. The negatively charged electron, which can be specified as e or e^-, participates in either an oxidation or reduction reaction. An oxidant is a substance that causes oxidation to occur while itself being reduced (accepting electrons), as follows:

$$\begin{array}{ll} O_2 + 4\,H^+ + 4\,e^- \rightarrow 2\,H_2O & \text{reduction} \\ \quad\quad\quad 4\,Fe^{2+} \rightarrow 4\,Fe^{3+} + 4\,e^- & \text{oxidation} \\ \hline O_2 + 4\,Fe^{2+} + 4\,H^+ \rightarrow 4\,Fe^{3+} + 2\,H_2O & \text{redox reaction} \end{array}$$

The classical description of redox potentials has involved the use of the term E_h, expressed in volts. The concept of P_ε is now replacing the use of E_h. As pH $= -\log[H^+]$, redox intensity can be written as $P_\varepsilon = -\log[e^-]$. In a highly reducing solution, the tendency to donate electrons is high. This is indicated by a low P_ε. High P_ε indicates a tendency to be oxidized, i.e., to accept electrons. The classical method of expressing the redox E_h is related to P_ε through the formula $P_\varepsilon = E_h(\text{volts})/0.059$. The P_ε values, at 25°C and pH 7 of a number of important biological reactions,

Table 2.4
Redox Pairs Arranged in Order from the Strongest Reductants (Negative Potentials) to the Strongest Oxidants (Positive Potentials)[a]

Redox pair	P_ε	E_h(volts)
CO_2/CH_2O	-7.3	-0.43
N_2/NH_4^+	-6	-0.35
$CO_2/$acetate	-4.7	-0.28
SO_4^{2-}/H_2S	-3.7	-0.22
Fumarate/succinate	$+0.51$	$+0.03$
NO_3^-/NO_2^-	7.11	0.42
MnO_2/Mn^{2+}	8.1	0.48
$Cyt_3(ox)/Cyt_3(red)$[b]	9.3	0.55
NO_3^-/N_2	12.4	0.74
Fe^{3+}/Fe^{2+}	12.9	0.76
$½\,O_2/H_2O$	13.9	0.82

[a]Some of the reactions shown involve several intermediate redox reactions.
[b]Cyt_3, cytochrome 3.

ranging from photosynthesis to decomposition, are given in Table 2.4. The greater the difference between electron potentials between donor and final acceptor, the greater the potential for energy capture by biological systems.

Soil pH

Many of the soils of the world are affected by excess acidity, a problem exacerbated by heavy fertilization with certain nutrients and by acid rain. Biological nitrogen fixation also creates acidity in that H^+ is produced during the fixation process. Measurements of pH are important criteria for predicting the capability of soils to support microbial reactions. Achieving a pH measurement of the soil solution is easy. However, negatively charged clays have a layer of positively charged ions (cations) attracted to them. Because of the increased concentration of cations within the double layer surrounding both clays and electronegative organic particles, the pH at the charged surface may be several times more acidic than that of the adjacent soil solution.

The biological transformation of NH_4^+ to NO_3^- (nitrification) is one of the most pH-sensitive soil reactions. The concept of pH differences attributable to the double-layer theory has been used to explain why optimum and minimum pH values for nitrification are different in soil and in laboratory solutions. Nitrification in forest soils can occur at measured pH values below 4, whereas nitrification has not been found to occur below pH 6 in solution culture. Part of this enigma may be explained by the occurrence in soil of microsites in which there is decomposition of nitrogen-rich materials. The release of ammonia creates a pH in the microsite higher than that in the soil solution. An alternative and more widely accepted explanation is that nitrification in acidic soils is caused by heterotrophic nitrifiers that are more acid tolerant than are the autotrophic nitrifiers.

The concept of pH values at a specific site must be related to the size of the organism and the multiplicity of enzymes at the microbial level. A bacterial cell contains about 1000 enzymes; many of these are pH dependent and associated with cell components, such as membranes. The pH optimum of enzymes is affected by absorption phenomena. In the soil matrix, adsorption of enzymes to the soil humates moves their pH optima to higher values. Until a better understanding and measurement of boundary and molecular pH values can be achieved, the soil microbiologist must be satisfied with the descriptive pH obtained by the traditional soil-paste measurement, which involves the addition of a $CaCl_2$ solution and measurement of the solution pH with an appropriate electrode.

Soil Temperature

Temperature affects not only the physiological reaction rates of cells but also most of the physicochemical characteristics of the environment; examples include soil volume, pressure, oxidation–reduction potentials, diffusion, Brownian movement, viscosity, surface tension, and water structure. The activities of microbial cells, as those of other organisms, are governed by the laws of thermodynamics. It is therefore not surprising that changes in soil temperature have marked effects on microbial activity.

The rate of a chemical reaction is a direct function of temperature and generally obeys the relationship originally described by Arrhenius:

$$k = Ae^{-E/RT}$$

where k is the reaction velocity, A the frequency with which molecules collide, E the activation energy of the reaction, R the gas constant, e the base of the natural logarithm, and T the temperature in kelvins (K). This can be rewritten as

$$\frac{k_2}{k_1} = \frac{\Delta E}{R}(\frac{1}{t_2} - \frac{1}{t_1})$$

At moderate temperatures, a plot of the reaction rate constant versus the reciprocal of the temperature in kelvins will yield a straight line with a slope E. This slope represents the energy hump (activation energy) that must be overcome for a reaction to proceed. Figure 2.8 shows the effect of temperature and pH on the rate constants for NH_4^+ oxidation by *Nitrosomonas*, and NO_2^- oxidation by *Nitrobacter*. Extremes of temperature cause a sharp fall-off at both ends of the curve. The abrupt fall in growth rate at high temperatures is caused by the thermal denaturation of proteins and alterations in the permeability of membranes. The maximum temperature of growth is the temperature at which these destructive forces become overwhelming. This temperature is usually only a few kelvins higher than the temperature at which the growth rate is maximal. The interaction of temperature response with pH, as shown in Fig. 2.8, explains why *Nitrobacter*, which oxidizes NO_2^-, is so much more sensitive to environmental conditions than *Nitrosomonas*, which oxidizes NH_4^+. *Nitrosomonas* shows a broad, linear response to temperature at all pH values. *Nitrobacter* on the other hand has a broad, linear response only at a pH of 7.3; at all other pH values it is very sensitive to changes in temperature.

The mechanism of heat tolerance is shown by studies on the kinetics of thermal denaturation of both enzymes and cell structures. The specific proteins, such as those of flagella and ribosomes, of thermophilic bacteria are more heat stable than those of mesophilic bacteria.

Figure 2.8. Combined effect of pH and temperature on reaction rate constants (hr^{-1}): ammonia oxidation (k_1) by *Nitrosomonas* (top graph); nitrite oxidation (k_2) by *Nitrobacter* (bottom graph). (From Wong-Chong and Loehr, 1978.)

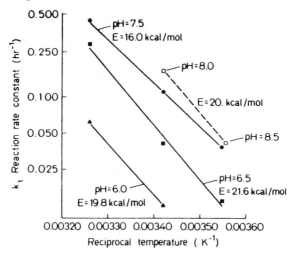

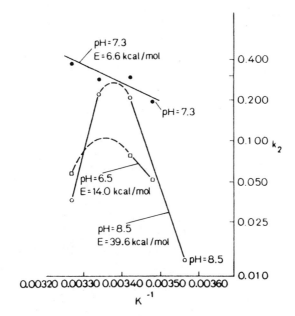

The factors controlling an organism's ability to operate at low temperatures also are related to cell structures. At low temperatures all proteins undergo slight conformational changes attributable to the weakening of the bonds that control tertiary (three-dimensional) structure. Since the shape of proteins is most important in materials, such as ribosomes, that are responsible for the genetic coding of proteins, mutations that increase the cold hardiness of cells usually occur in genes coding for protein production. The melting point of lipids is directly related to their content of saturated fatty acids. Consequently, the degree of saturation of fatty acids in membrane lipids determines their degree of fluidity at given temperatures. Since membrane function depends on fluidity, it follows that growth at low temperatures is facilitated by an increase in the degree of unsaturation of fatty acids.

Most measurements relating microbial activity to temperature show growth stopping at 0°C. Some psychrophilic bacteria are capable of growth below the freezing point, providing the osmotic concentration of the ambient solution or of the organism's cytoplasmic constituents is sufficiently high to permit the cell interiors to remain unfrozen. A generalized temperature response curve for microbial activity is shown in Fig. 2.9. There is a fairly sharp alteration in response at 10°C and a flat response between 25 and 35°C. Individual species differ in their optimum temperature, but the general shape of temperature response curves is quite similar for many organisms. For a given organism, the minimum temperature for continuation of growth once started may be somewhat lower than the minimum

Figure 2.9.   Relative microbial reaction rates at various temperatures.

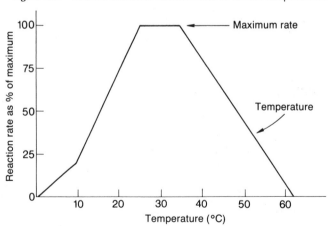

for initiation of growth. Thus at a low temperature that is borderline for initiation of growth, microorganisms may show growth if supplied as a large inoculum but fail to do so if only a small inoculum is used.

Very few soils maintain a uniform temperature in their upper layers. Variations may be either seasonal or diurnal. Because of the high specific heat of water, wet soils are less subject to large diurnal changes than dry soils. Among factors affecting the intensity and reflectance of solar irradiation are the soil aspect (south or north slopes, etc.), steepness of slope, degree of shading, and surface cover (vegetation, litter, mulches). Diurnal changes level out with profile depth. Measurements for a given site in midsummer have shown diurnal fluctuations of 15 to 18°C, 8 to 10°C, and 1 to 2°C for soil depths of 5, 10, and 30 cm, respectively.

Interactions of Environmental Factors

It is difficult to interpret the interactions involving temperature, moisture, soil pH, soil aeration, redox potential, and soil type. In nature, stress factors very seldom act independently. Only computer modeling techniques can attempt to describe all the interactions. A semi-perspective plot of computed respiration rate at different temperature and moisture levels is shown in Fig. 2.10. A summary of microbial adaptations to a number of stress factors is given in Table 2.5.

Figure 2.10.   Semiperspective plot of computed respiration rate at different temperature and moisture levels. All curves actually begin at origin. (From Bunnell and Tait, 1974.)

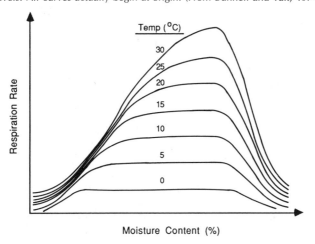

Table 2.5
Effects of Stress Conditions on Microorganisms and Biochemical Adaptations Induced

Stress	Effect on cells	Organisms involved	Biochemical Adaptations or responses
Heat	Denaturation of enzymes	Thermophiles	Synthesis of heat-stable proteins
Cold	Decrease in membrane fluidity	Psychrophiles	Production of more unsaturated fatty acids
Water potential	Dehydration and inhibition of enzyme activity	Osmophiles, halophiles, xerophiles	Compensating, solute accumulation, enzyme adaptations
Acidity	Protein denaturation, enzyme inhibition	Acidophiles	Proton exclusion, adaptations in surface appendages
Anaerobiosis	Alteration of metabolic pathway	Anaerobes, microaerophiles	Use of alternate electron sinks, fermentation

References

Bunnell, F. L., and Tait, D. E. N. (1974). Mathematical simulation models of decomposition processes. *In* "Soil Organisms and Decomposition in Tundra" (A. J. Holding *et al.*, eds.), pp. 297–324. Tundra Steering Committee, Stockholm.

Campbell, R., and Rovira, A. D. (1973). The study of the rhizosphere by scanning electron microscopy. *Soil Biol. Biochem.* **5,** 747–752.

Chen, Y., and Avnimelech, Y. (1986). "The Role of Organic Matter in Modern Agriculture." Martinus Nijhoff Publ., The Hague.

Focht, D. D., and Martin, J. P. (1979). Microbiological and biochemical aspects of semi-arid agricultural soils. *Ecol. Stud.* **34,** 119–147.

Harris, R. F. (1981). Effect of water potential on microbial growth and activity. *In* "Water Potential Relation in Soil Microbiology" (J. Parr *et al.*, eds.). Soil Sci. Soc. Am., Madison, Wisconsin.

Lynch, J. M., and Poole, N. J. (1979). "Microbial Ecology: A Conceptual Approach." Wiley, New York.

McLaren, A. D., and Skujins, J. (1968). The physical environment of microorganisms in soil. *In* "The Ecology of Soil Bacteria" (T. R. G. Gray and D. Parkinson, eds.), pp. 3–24. Univ. of Toronto Press, Toronto.

Stumm, W., and Morgan, J. J. (1981). "Aquatic Chemistry: An Introduction Emphasizing Chemical Equilibria in Natural Waters," 2nd ed. Wiley, New York.

Tiedje, J. M., Sextone, A. J., Parkin, T. B., Revsbach, N. P., and Shelton, D. R. (1984). Anaerobic processes in soil. *In Dev. Plant Soil Sci.* **2,** 197–212.

Tisdall, J. M., and Oades, J. M. (1982). Organic matter and water stable aggregates in soil. *J. Soil Sci.* **32,** 141–163.

Wong-Chong, G. M., and Loehr, R. C. (1978). Kinetics of microbial nitrite nitrogen oxidation. *Water Res.* **12**, 605–609.

Supplemental Reading

Brock, T. D., Smith, D. W., and Madigan, M. T. (1984). "Biology of Microorganisms," 4th ed. Prentice-Hall, Englewood Cliffs, New Jersey.

Campbell, R. (1977). "Microbial Ecology." Blackwell, Oxford.

de Jong, E., and Paul, E. A. (1979). Soil aeration. *In* "Encyclopedia of Soil Science" (R. Fairbridge and C. W. Finkl, Jr., eds.), pp. 10–19. Dowden, Hutchinson & Ross, Stroudsburg, Pennsylvania.

Emerson, W. W. (1977). Physical properties and structure. *In* "Soil Factors in Crop Production in Semiarid Environment" (J. S. Russell and E. L. Greacen, eds.). Univ. of Queensland Press, St. Lucia.

Griffin, D. M. (1972). The influence of water. *In* "Ecology of Soil Fungi" (D. M. Griffin, ed.), Chapter 5. Syracuse Univ. Press, Syracuse, New York.

Griffin, D. M. (1981). Water and microbial stress. *In* "Advances in Microbial Ecology" (M. Alexander, ed.), pp. 91–136. Plenum, New York.

Gupta, S. C., Radke, K., and Larson, W. E. (1981). Predicting temperatures of bare and residue covered soils without a corn crop. *Soil Sci. Soc. Am. J.* **45**, 405.

Ingraham, J. L., Maale, O., and Neidhart, F. C. (1981). "Growth of the Bacterial Cell." Sinuaer Assoc., Sunderland, Massachusetts.

Lynch, J. M. (1983). "Soil Biotechnology, Microbial Factors in Crop Productivity." Blackwell, Oxford.

Russell, E. W. (1973). "Soil Conditions and Plant Growth," 10th ed. Longmans, Green, London.

Soil Science Society American Terminology Committee (1984). "Glossary of Soil Science Terms," rev. ed. Soil Sci. Soc. Am., Madison, Wisconsin.

Stotsky, G. (1974). Activity, ecology, and population dynamics of microorganisms in soil. *In* "Microbial Ecology" (A. Laskin and H. Lechevalier, eds.), pp. 57–135. CRC Press, Cleveland, Ohio.

Chapter 3

Methods for Studying
Soil Organisms

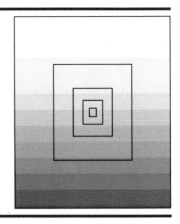

Introduction

A soil sample taken for microbiological study is a composite of a large number of microhabitats occurring in nature. One needs the test tube to unravel specific functions, but the real purpose of the agronomist or ecologist is to determine what is happening in the field site being studied. To achieve that, one must operate at numerous levels, including molecular, microbiological, whole-plant, and field measurements.

Approaches for estimating the kinds, numbers, and metabolic activities of organisms in soil and plant–microbial associations are shown in Fig. 3.1 and include determination of the form and arrangement of microorganisms in soil, isolation and characterization of subgroups and species, and detection and measurement of metabolic processes. Collectively, these provide a general estimate of the numbers and types of organisms in soil, their biomass, and the functions they carry out. Detailed descriptions of procedures for studying soil organisms have been given by Gerhardt *et al.* (1981), Page (1982), and Parkinson *et al.* (1972).

Collection of Soil Samples

Analytical data obtained from soil samples become more informative if supplemented by information concerning the sample site. The following physical, chemical, and biotic factors are of value:

Topography	Particle size and type	Plant cover and productivity
Parent material	CO_2 and O_2 status	Vegetation history

Soil type	Chemical analysis	Management history
Moisture status	Temperature: range and variation	Animal presence
Soil pH	Rainfall: amount and distribution	Organic matter inputs and roots present

Obtaining a representative sampling that can be treated statistically is important. A knowledge of the field usually suggests that low or high spots,

Figure 3.1. Methods for determining microbial biomass and activity.

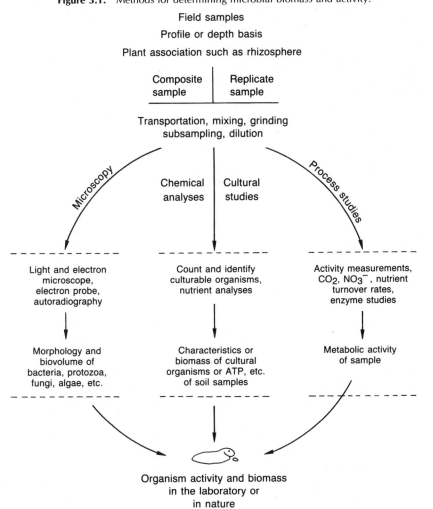

which have different soil characteristics and quite probably different nutrient content, should be sampled separately. In fairly uniform areas, a minimum of four replicates, each made up of at least two separate collections, is required to reduce the error to about 10% of the measurement. Geostatistics, or kriging is designed to handle field data where observations are not spatially independent. This can better account for field variability and often is much more useful than classical statistics (Marx and Thompson, 1987). For nutrient balance work, such as with nitrogen-15, random subsampling seldom gives adequate balances. Cylinders or open lysimeters allowing complete mixing of the total enclosed soil or of specific layers before subsampling are necessary.

For soil samples that are to be used for chemical analyses, immediate freezing or drying as soon as possible is generally satisfactory. There are exceptions; for example, in measuring soil nitrite, drying is not desirable as it is destructive of nitrite. For nitrate analysis, the soil should be dried within 24 hr, otherwise the nitrate level is subject to increase before or during the drying process. For biological analysis, transportation and storage time should be kept to a minimum. Storing the moist soil at either the field temperature or 4°C is employed when samples cannot be immediately processed. The soil disturbance incidental to sampling may in itself trigger changes in the soil population during the storage interval. Thus observations on a stored sample may not be representative of the undisturbed field soil.

Direct Microscopy of Soil

Direct examination is especially useful for determining the form and arrangement of microorganisms in soil. Thin-section techniques, as used in soil micromorphology with normal light microscopy, permit observations of soil structure, root penetration, and the soil fauna. Thin sections are usually not thin enough to permit observations of the soil microflora. Fluorescent stains, such as magnesium sulfonic acid, acridine orange, and fluorescein isothiocyanate, make it possible to observe microorganisms on soil particles, providing the light is focused on the surface of the particle.

Thin sections and the use of transmission electron microscopy (TEM) have proven useful, for studies on soil fabric–microbial interactions and for observing microorganisms on root segments (Fig. 3.2). The complex shadowing technique used in TEM makes it difficult to distinguish microorganisms from soil particles. Scanning electron microscopy (SEM), which gives a three-dimensional view, has proved superior for showing both the shape and structure of the microbial habitat and the microor-

Figure 3.2. A thin section of the root—rhizosphere area of wheat. The root in the lower portion of the picture has a colony of bacteria on its surface. The adjacent soil has one or two bacteria. (Photo courtesy of R. Foster.)

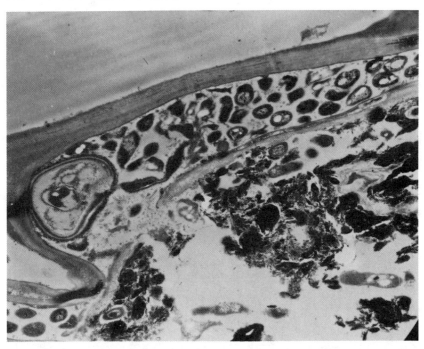

ganisms themselves. Bacteria are often enclosed in slime and are difficult to see, but fungal and actinomycete structures are easily observed (Fig. 3.3). SEM is basically a small probe of electrons that move rapidly across a soil surface. The back-scatter of these electrons gives the three-dimensional effect. By measuring back-scattered electrons at specific wavelengths or by using an electron probe that can be focused into very closely defined wavelengths, it is possible to analyze for individual nutrients. Thus *in situ* analyses of roots and soil organisms is now a possibility, although some instrumental problems still exist.

The buried-slide (Ross–Cholodny) technique entails burial of glass slides in soil for extended periods. Microorganisms adhere to the slide, and on its removal and staining, the qualitative nature of the soil population can be observed. Interrelationships among various groups of organisms and types of colony formation are readily seen on slides containing no added nutrients. The addition of a clear coating, such as cellophane, makes it possible to observe the sequence of colonization of a complex substrate.

Figure 3.3. Scanning electron micrograph of decomposing leaf litter. The bacteria tend to be hidden by slime but various sized filamentous organisms are readily apparent (photo courtesy of R. Todd.)

Some information on the arrangement of microorganisms can be obtained by coating a glass slide with a gluing agent and touching a soil exposure with the sticky surface, followed by microscopic examination.

The microorganisms in soil grow in pores or within aggregates, and the relationships seen on the surface of a flat slide may not be fully representative. Russian workers have met this difficulty by using pedoscopes. These are thin, optically flat capillary tubes that can be buried in the soil, with or without added nutrients coating the inside walls and left undisturbed for periods long enough to allow steady-state conditions to become established. A great deal of effort is involved in preparing optically flat capillaries, and the technique is not widely used. Other workers utilize fine nylon mesh buried in soil. The gauze pieces on removal are mounted and stained. Fungal development is followed by counting the number of openings filled by fungi after different times of burial.

Population Counting by Direct Microscopy

Direct microscopy of a soil suspension is a common procedure for counting soil microorganisms. The soil is dispersed with agents such as sodium pyrophosphate or detergents (Tween 80) in a Waring blender to disrupt soil aggregates. A suitable volume of the suspension is then either placed on a 1-cm^2 area of a microscopic slide or mixed with agar in a microbial counting chamber of known dimensions. After the agar hardens, it is floated off the counting chamber and placed on a microscopic slide for fixing and staining. This technique, known as the Jones and Mollison slide technique, provides a known dilution of soil particles and microorganisms in a known volume. The method is especially useful for fungi; bacteria are more difficult to count as they are not readily distinguishable from soil particles. A technique having a long history in water studies is to filter a soil suspension through either a colorless or black polycarbonate filter. The filter, plus its associated microorganisms, is then mounted on a glass slide and stained. Organisms strongly adsorbed on the solid matrix or entrapped in aggregates fail to enter the aqueous phase of the soil suspension.

There are a number of staining techniques. The traditional rose bengal and phenolic aniline blue stains have largely been replaced by fluorescent stains. Acridine orange, the oldest and most widely used, stains the DNA inside the cells. Living cells take up small quantities of the dye and appear green; dead cells lacking a functional cytoplasmic membrane absorb large quantities of the dye and appear red. However, the color of the cells is also dependent on the concentration of the dye, the nature of the cell wall, and the ratio of RNA to DNA within the cytoplasm. Clear differentiation of living versus dead is difficult. Acridine orange is adsorbed to soil particles and does not give a good stain in clay soils, but it is useful for sandy soils.

Fluorescein isothiocyanate (FITC) adsorbs to the sulfydryl groups in protein and is excellent for bacteria but does not adsorb to all fungi. For fungi, a fluorescent dye, water-soluble aniline blue, which is preferentially adsorbed to the $\beta(1\rightarrow3)$-glycan linkages in fungal cells is suitable. The use of fluorescein diacetate (FDA) as a stain should in theory separate active from inactive hyphae. After absorption, the FDA, initially nonflu-

Table 3.1
Equations for Calculating Biomass Values

Calculation of bacterial numbers in soil:

$$N_g = N_f \frac{A}{a_m} \frac{V_{sm}}{V_{sa}} D \frac{W_w}{W_d}$$

N_g ≡ number of bacteria per gram dry soil
N_f ≡ bacteria per field
A ≡ area (mm²) of smear (or filter)
A_m ≡ area (mm²) of microscope field
V_{sm} ≡ volume (ml) of smear of filter
V_{sa} ≡ volume (ml) of sample
D ≡ dilution
W_w ≡ wet weight soil
W_d ≡ dry weight soil

Bacterial biomass as carbon:

$$C_b = N_g V_b e S_c \frac{\%C}{100} \times 10^{-6}$$

c_b ≡ bacterial biomass carbon (μg g^{-1} soil)
N_g ≡ number of bacteria per gram soil
v_b ≡ average volume (μm³) of bacteria ($\pi r^2 L$; r ≡ radius, L ≡ length)
e ≡ density (1.1 × 10^{-3} in liquid culture)
S_c ≡ solids content (0.2 in liquid culture, 0.3 in soil)
$\%C$ = carbon content (45% dry weight)

Calculation of fungal biomass carbon:

$$C_f = \pi r^2 L e S_c \%C \times 10^{10}$$

C_f ≡ fungal carbon (μg carbon g^{-1} soil)
r ≡ hyphal radius (often 1.13 μm)
L ≡ hyphal length (cm g^{-1} soil)
e ≡ density (1.1 in liquid culture, 1.3 in soil)
s_c ≡ solids content (0.2 in liquid culture, 0.25–0.35 in soil)

orescent, is cleaved by an internal esterase to fluorescein and acetate. The free fluorescein is capable of fluorescence. When used on soil preparations, fluorescence is shown by 2 to 10% of fungal hyphae. It is a good indicator of young hyphae and hyphal tips.

Calculation of Biovolume and Biomass

The microbial biovolume is determined from the diameter and length of the organisms. For rod-shaped organisms, the ends are assumed to be a hemisphere attached to each end of a cylinder. The biomass of the organisms on a dry-weight basis can be determined as shown by equations given in Table 3.1. Representative soil fungi, grown at moisture tensions ranging from -0.03 to -1.3 MPa, have been found to be 50% heavier in weight than when grown in laboratory shake cultures. This indicates that volume-to-biomass conversion should be 0.33 rather than the often quoted value of 0.22. Table 3.1 shows that on a laboratory culture basis, the ratio of biovolume to biomass for bacteria is also 0.22. Measurements on bacteria grown under conditions similar to those in soil also often show values twice as great as this.

Direct microscopy and fluorescent staining can be utilized for determining the number and biovolume of specific organisms in soil. This is done with immunofluorescent techniques (Fig. 3.4). Specific antigens,

Figure 3.4. Utilization of direct and indirect antibody staining techniques. Fluorescent isothiocyanate is usually used to label the organism; radiotracers or enzyme-linked immunosorbants can also be employed.

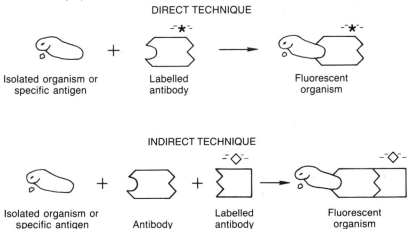

DIRECT TECHNIQUE

Isolated organism or
specific antigen   Labelled
antibody

Fluorescent
organism

INDIRECT TECHNIQUE

Isolated organism or
specific antigen   Antibody   Labelled
antibody

Fluorescent
organism

which may be whole cells or materials isolated from them, are injected intravenously into rabbits. The antiserum produced is collected and labeled with a fluorescent stain, such as FITC. Reaction of serum with a mixture of unknown organisms on a microscopic slide or filter results in the labeling of the specific organism with a green fluorescence. Using this technique, specific counts of *Aspergillus flavus, Rhizobium* spp., and nitrifying bacteria in soil and on root systems have been obtained. Problems with nonspecific reactions and the need to work with characterized organisms have limited the application of this technique to soils.

Attaching a specific enzyme, such as phosphatase, rather than FITC to the antibody broadens the scope of the technique and makes it easier to use. These analyses by enzyme-linked immunosorbants (ELISA) have been found useful in identification of rhizobia strains in soil.

Biomass Measurement by Chemical Techniques

Soil organisms subjected to chloroform ($CHCl_3$) vapor have their cell membranes destroyed by the vapor. This allows cell constituents to leak into the soil. The cytoplasmic phosphorus can be determined directly on an appropriate extract, as can some of the carbon, nitrogen, and sulfur constituents. However, biomass carbon and nitrogen determinations usually involve the incubation of the soil containing the lysed cells after inoculation with a small amount of soil having a living population. Degradation of the freshly killed population over a 10-day period yields CO_2 and NH_4^+. The relationship between the amount of CO_2 or NH_4^+ evolved and biomass is determined separately with ^{14}C- or ^{15}N-labeled cells added to the soil and then subjected to $CHCl_3$. Although the amount of carbon that is evolved varies for fungi and bacteria, an average 41% of fungal and bacterial C is evolved as CO_2. Biomass is calculated as follows:

$$B_C = F_C/K_C \qquad \text{or} \qquad B_N = F_N/K_N$$

where B_C is the biomass of carbon, B_N the biomass of nitrogen, F_C the flush of decomposition (i.e., CO_2 from $CHCl_3$-treated soils, or CO_2 from treated minus soil-corrected CO_2 from untreated soil), and K_C the percentage biomass of carbon mineralized to CO_2 (equals 0.41 for many soils).

The amount of nitrogen mineralized after $CHCl_3$ vapor treatment and incubation varies with the carbon:nitrogen ratio of the organisms. Calculation of a factor K_N to take into account the reimmobilization of nitrogen after taking fumigation into account has been found to result in the equation

$$K_N = 0.8(C_F/N_F)^{-0.43}$$

where K_N is the percentage of nitrogen mineralized after fumigation, C_F the flush of CO_2 carbon, as defined above, and N_F the flush of NH_4^+ nitrogen. The problems with reimmobilization of the nitrogen and the nuisance of a 10-day incubation can be by-passed by the direct determination of carbon and nitrogen extractable after $CHCl_3$ treatment (Vance *et al.,* 1967; Amato and Ladd, 1988).

The $CHCl_3$ technique for biomass is simpler than direct microscopy. It permits measurement of tracers such as carbon-14 and nitrogen-15 incorporated into the soil organisms. Methods for sieving fungi from soils and for separating bacteria in density gradient solutions also now make it possible to isolate microorganisms from soil directly.

Measurement of ATP

All biosynthetic and catabolic reactions within cells require the participation of adenosine triphosphate (ATP) which is a derivative of adenosine monophosphate formed by the addition of two high-energy phosphate bonds. Thus ATP is able to donate phosphate groups to other metabolic intermediates, converting them to activated forms.

ATP is common to all life and is very sensitive to environmental factors and phosphatases. It does not persist in soil in a free state. Consequently, this coenzyme should be an ideal material for determining the amount or activity of life within soil, sediment, or aquatic systems. ATP is a central reactant in many reactions, many of which theoretically could be used for its measurement. One of the most straightforward is the use of the luciferin reaction, as shown in Fig. 3.5. The substrate luciferin, an aromatic nitrogen- and sulfur-containing molecule, can react with ATP and luciferase in the presence of magnesium to give an enzyme–luciferin–adenosine monophosphate intermediate. This, in the presence of O_2, breaks down to produce free adenosine monophosphate, inorganic phosphorus, and light. There is very sharp emission of light during the first 15 sec of the reaction, after which emission continues at a fairly steady state for 1 to 3 min. The light emitted is measured by a photometer or scintillation counter and plotted against ATP content to form a standard curve. If pure luciferase and luciferin rather than firefly tails are used, the light output is extended and constant.

ATP is extracted from microbial cultures by lysing the cells. In pure culture this can be done by boiling, by acids (e.g., H_2SO_4, $HClO_4$), or by $CHCl_3$. In soils and sediments, boiling does not give adequate lysis, indicating that the cells are protected by the clay colloids. Chloroform in 0.5 M $NaHCO_3$ has been found useful in calcareous soils, but the ATPases

Figure 3.5. Mechanism of light emission in ATP reaction with luciferin and the enzyme luciferase.

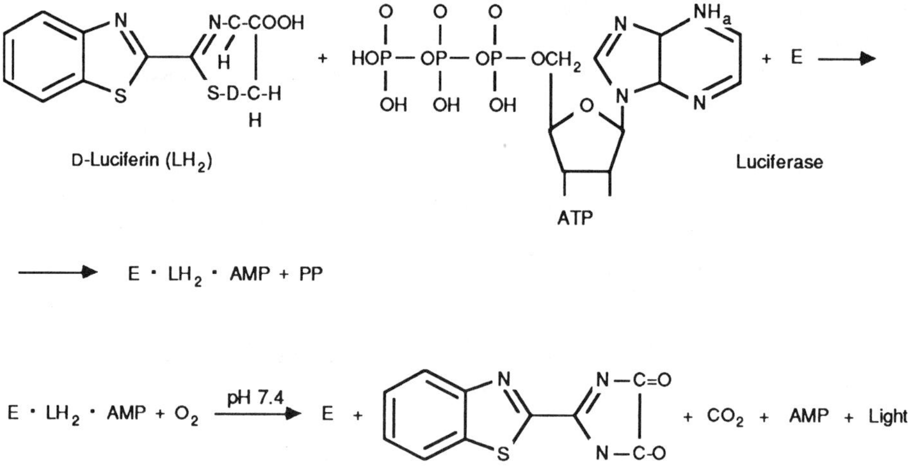

$$E \cdot LH_2 \cdot AMP + PP$$

of soil are not all inhibited and this leads to low results. Extracting reagents found useful include combinations of anions (phosphate) and cations (Paraquat). Most extraction reagents inhibit the luciferase reaction to some extent; therefore, the solution is diluted as much as possible before being placed in the ATP photometer.

During microbial growth, the carbon:ATP ratio can vary from 1000:1 to 40:1. In the resting state, the ATP:cell nutrient ratios are usually as shown:

ATP	:	C	:	N	:	P	:	S
1	:	250	:	40	:	9	:	2.6

The biomass carbon contents in a number of soils, as measured by $CHCl_3$, ATP, and direct microscopy, are shown in Table 3.2. ATP is a measure of both biomass and activity and is influenced by the soil phosphorus content. It can be used most successfully to characterize soils whose microbial population is in the resting state at excess or constant phosphorus levels.

Measurement of Respiration

Microbial cells, in common with all life, oxidize reduced materials such as carbohydrates according to the generalized equation

$$CH_2O + O_2 \rightarrow CO_2 + H_2O + \text{intermediates} + \text{cellular material} + \text{energy}$$

Table 3.2
Estimates of Soil Microbial Biomass by Different Methods[a]

Field source	Biomass carbon $\mu g\ g^{-1}$ soil)		
	Chloroform fumigation	Direct count	ATP method
Continuous wheat plus manure, England	560	500	430
Continuous wheat, no manure, England	220	170	170
Calcareous deciduous wood, England	1230	1400	1040
Acid deciduous wood, England	50	300	470
Old grassland, England	3710	2910	—
Secondary rain forest, Nigeria	540	390	—

[a]From Jenkinson and Ladd, (1981).

Under anaerobic conditions, the most common heterotrophic reaction is that of fermentation:

$$C_6H_{12}O_6 \rightarrow 2\ CH_3CH_2OH + 2\ CO_2 + energy$$

or methane production:

$$H_2 + CO_2 \rightarrow CH_4 + H_2O + energy$$

The mechanisms of metabolism are covered in general microbiology and biochemistry texts such as those by Stanier et al. (1986), Lehninger (1982), Dawes and Sutherland (1976), Atlas (1984), and Brock et al. (1984).

Fermentation occurring within aggregates or under waterlogged conditions produces soluble intermediates that either accumulate or diffuse to aerobic areas and are there transformed to CO_2 and O_2. Respiratory measurement of microbial activity can determine either CO_2 or O_2. Because the atmospheric CO_2 concentration is only 0.035%, versus 20% for O_2, measurements for CO_2 are more sensitive. Measurement of CO_2 also allows a direct balance of the carbon in growth relative to substrate decomposition to be made. Methods of CO_2 measurement include aeration trains, in which NaOH is used to trap evolved CO_2 in an air stream from which CO_2 is removed before the air is exposed to the soil sample. The reaction occurs as follows:

$$2\ NaOH + CO_2 \rightarrow Na_2CO_3 + H_2O$$

Before titration, $BaCl_2$ is added to precipitate the CO_3^{2-} as $BaCO_3$ and excess NaOH is back-titrated with acid. Each milliequivalent of NaOH used to absorb evolved CO_2 is equivalent to 6 mg of CO_2 carbon.

In the laboratory, NaOH containers placed in sealed jars are convenient and effective for CO_2 absorbence. The jar must be opened at intervals so that the O_2 concentration does not drop below 10%. Gas chromatography (GC), with thermal conductivity detectors, can be used for measuring CO_2 concentration after CO_2 is separated from other constituents on column materials such as Poropak Q. Computer-operated valves in conjunction with GC allow time-sequence studies to be handled automatically. Infrared gas analyzers are sensitive to CO_2 and can be used for both static and flow systems after H_2O, which adsorbs in the same general wavelength, has been removed.

In the field, canopies, either on top of the soil or surrounding a soil area, or tubes sunk into the soil at various depths are commonly employed to measure CO_2. With canopies, one must be aware of the possibility of CO_2 diffusion from beneath the canopy and the possibility of CO_2 flow through cracks. With tubes installed at various depths, knowledge of the CO_2 concentrations at the different depths can be utilized with diffusion equations to calculate the flux of CO_2 from the soil surface. The possibility of CO_2 carbon equilibrating with $CaCO_3$, of dissolved CO_2 in water, and of CO_2 production or utilization by plant roots must be taken into account when measuring CO_2 evolution as an index of microbial activity.

Oxygen analyses usually involve manometers in a system where CO_2 is absorbed by NaOH. Polarographic and paramagnetic O_2 analyzers are also used but are not as sensitive or as stable as CO_2-measuring devices.

The measurement of CO_2 can be augmented by the incorporation of carbon-14 into chosen substrates (radiorespirometry). The carbon-14 may be in known molecules, such as glucose, cellulose, amino acids, or herbicides, or in such complex materials as microbial cells or plant residues. The oxidation of organic substrates is the major source of energy for heterotrophic growth, and all aerobic degradations involve the evolution of CO_2. Radiorespirometry can be used in following microbial activity during the degradation of plant and animal residues and in determining the fate of manmade chemicals added to the environment. The use of these techniques in plant residue decomposition, microbial growth, soil organic formation, and the geocycle of carbon are discussed later.

Enzymes and Their Measurement

Life is composed of a series of enzyme reactions, and enzymes carry out most of the reactions in nutrient cycling. Not surprisingly, a great amount of work has been done on the measurement of soil enzymes; a listing of those commonly measured is given in Table 3.3. Some enzymes (e.g., urease) are constitutive and are routinely produced by cells; others are adaptive or induced, being formed only in the presence of a susceptible

Table 3.3
Some Soil Enzymes Found in Soil and the Reactions They Catalyze

Enzyme	Reaction catalyzed
Oxidoreductases	
Catalase	$2H_2O_2 \rightarrow 2H_2O + O_2$
Catechol oxidase (tyrosinase)	O-Diphenol $+ \frac{1}{2} O_2 \rightarrow O$-quione $+ H_2O$
Dehydrogenase	$XH_2 + A \rightarrow X + AH_2$
Diphenol oxidase	P-Diphenol $+ \frac{1}{2} O_2 \rightarrow P$-quinone $+ H_2O$
Glucose oxidase	Glucose $+ O_2 \rightarrow$ gluconic acid $+ H_2O_2$
Peroxidase and polyphenol oxidase	$A + H_2O_2 \rightarrow$ oxidized A $+ H_2O$
Transferases	
Transaminase	$R_1R_2\text{-CH-N}^+H_1 + R_1R_4CO \rightarrow R_1R_4\text{-CH-N}^+H_3 + R_1R_2CO$
Hydrolases	
Acetylesterase	Acetic ester $+ H_2O \rightarrow$ alcohol $+$ acetic acid
α- and β-Amylase	Hydrolysis of $\beta(1 \rightarrow 4)$ glucosidic bonds
Asparaginase	Asparagine $+ H_2O \rightarrow$ aspartate $+ NH_1$
Cellulase	Hydrolysis of $\beta (1 \rightarrow 4)$ glucan bonds
Deamidase	Carboxylic acid amide $+ H_2O \rightarrow$ carboxylic acid $+ NH_3$
α- and β-Galactosidase	Galactoside $+ H_2O \rightarrow$ ROH $+$ galactose
α- and β-Glucosidase	Glucoside $+ H_2O \rightarrow$ ROH $+$ glucose
Lipase	Triglyceride $+ 3 H_2O \rightarrow$ glycerol $+ 3$ fatty acids
Metaphosphatase	Metaphosphate $\rightarrow$ orthophosphate
Nucleotidase	Dephosphorylation of nucleotides
Phosphatase	Phosphate ester $+ H_2O \rightarrow$ ROH $+$ phosphate
Phytase	Inositol hexaphosphate $+ 6 H_2O \rightarrow$ inositol $+ 6$-phosphate
Protease	Proteins $\rightarrow$ peptides $+$ amino acids
Pyrophosphatase	Pyrophosphate $+ H_2O \rightarrow 2$-orthophosphate
Urease	Urea $\rightarrow 2 NH_1 + CO_2$

substrate or some other initiator. An example of an induced enzyme is cellulase; it is produced in the presence of cellulose. Dehydrogenases are often measured because they are constitutive and found only in living systems. Various dyes, e.g., triphenyltetrazolium chloride, which is colorless in the oxidized state but red in the reduced state, can be added to soil and extracted in the reduced form as a measure of dehydrogenase activity of soil.

Soil enzymes are proteins and are often entrapped in soil organic and inorganic colloids. Therefore, the soil has a large background of extracellular enzymes not directly associated with the microbial biomass.

Cultural Counts of Microorganisms

Culturing of a soil organism involves transfer of its propagules to a substrate conducive to growth. All cultural techniques are selective and designed to detect microorganisms with particular growth forms or biochemical capabilities. Most techniques depend on the use of soil suspensions, with the degree of dilution required being related to the initial number of organisms in the soil. Serial dilutions, usually one-tenth, stepwise, are made on a known weight of wet or dry soil. Aggregates are broken in the Waring blender, often in the presence of a dispersing agent such as $Na_4P_2O_7$. Replicate 1-ml portions of appropriate suspensions are transferred to solid or liquid substrates on which individual propagules develop into visible growths. Solid substrates are used for plate counts, and liquid substrates for the extinction dilution count.

In plate counts, 1 ml of inoculum is placed in each of three or five Petri dishes and mixed with melted agar (usually about 20 ml) at 50°C. Alternately, agar can be prepoured into plates and the inoculum spread over the surface of the hardened agar. After appropriate incubation, single colonies are counted, and each colony is equated with a single propagule in the soil suspension. Mycologists usually prefer to use the term cfu (colony-forming unit) in lieu of propagule.

A wide variety of media and incubation conditions can be used. Soil extract agar, prepared by boiling 1 liter of water mixed with 1 kg of soil, followed by centrifugation, or filtration of the supernatant and solidification of the extract with 1.5% agar, yields the largest number of colonies. This medium contains the wide range of the amino acids and growth factors required by soil microorganisms, but only a very low carbon supply. Consequently, colonies are slow growing and usually very small. Nutrient agar containing meat extract ingredients yields nearly as large numbers. The higher available energy in nutrient agar also leads to larger colonies, easier counting, and often, pigment production. Media adjustments for different groups of organisms include the use of antibiotics, such as streptomycin, to eliminate bacteria but not fungi. Actidione is used to eliminate fungi but not bacteria. Other modifications include acidification to favor fungi, certain dyes to favor Gram-negative bacteria, and diverse selective media for groups such as actinomycetes, nitrogen fixers, or sulfur oxidizers. The plate count of bacteria usually represents only 1–5%, and at the most, 50%, of the number determined by direct microscopy.

Where plate counts are not appropriate, dilution counts can be made. The dilution count, in essence, is a determination of the highest soil dilution that will still provide growth in a suitable medium. Employment of replicate inocula (10 usually, 5 minimum) from each of three successive serial dilutions at the estimated extinction boundary for growth enables resultant visual growth to be converted to numbers within statistical limits of reliability. For conversion tables, see Cochran (1950), Alexander (1982), or Gerhardt *et al.* (1981).

Isolation of specific microorganisms is usually accomplished culturally. The culture medium employed is tailored to the individual species desired. Isolations may also be made microscopically. Single cells, spores, or hyphal fragments may be picked out on a microscopic stage equipped with micromanipulators. Plants may be used to trap specific microorganisms. The nodules formed on susceptible legume plants provide a source for isolation of rhizobia. Planting a susceptible host permits screening of a mixed soil population for a root-invading pathogen. The use of antibiotic-resistant strains makes it possible to readily reisolate specific isolates that have been inoculated into soil for process studies. DNA probes, which can identify specific genetic coding in cells (Pace *et al.*, 1986), have been successfully utilized on organisms cultured from soil. Direct detection of microbial DNA without subculturing presents a number of analytical problems but provides much useful information (Holben *et al.*, 1988).

References

Alexander, M. (1982). Most probable number method for microbial populations. *Agron. Monogr.* **9**, 815–820.

Amato, M., and Ladd., J.N. (1988). Assay for microbial biomass based on ninhydrin-reactive nitrogen in extracts from fumigated soils. *Soil Biol. Biochem.* **20**, 107–114.

Atlas, R. M. (1984). "Microbiology, Fundamental and Application." Macmillan, New York.

Brock, T. C., Smith, D. W., and Madigan, M. T. (1984). "Biology of Microorganisms." Prentice-Hall, Englewood Cliffs, New Jersey.

Cochran, W. G. (1950). Estimation of bacterial densities by means of the "most probable number." *Biometrics* **6**, 105–116.

Dawes, W., and Sutherland, I. W. (1976). "Microbial Physiology." Blackwell, Oxford.

Foster, R. C. (1985). *In situ* localization of organic matter in soils. *Quest. Entomol.* **21**, 609–633.

Gerhardt, P., Murray, R. G. E., Costilow, R. N., Nestor, E. G., Wood, W. A., Kaeg, N. R., and Phillips, G. B. (1981). "Manual of Methods for General Bacteriology." Soc. Amer. Microbiol. Washington, D.C.

Holben, W.E., Jansson, J.K., Chelm, B.K., and Tiedje, J.M. (1988). DNA probe method for the detection of specific microorganisms in the soil bacterial community. *Appl. Env. Microbiol.* **54**, 703–711.

Jenkinson, D. S., and Ladd, J. N. (1981). Microbial biomass in soil: Measurement and turnover. *In* "Soil Biochemistry" (E. A. Paul and J. N. Ladd, eds.), vol. 5, pp. 415–471. Dekker, New York.

Lehninger, A. L. (1982). "Biochemistry." Worth Publishers, New York.

Marx, D., and Thompson, K. (1987). Practical aspects of agricultural kriging. Bull. 903. Arkansas Agr. Exp. Sta. Fayetteville, Arkansas.

Pace, N.R., Stahl, D.A., Lane, D., and Olsen, G.J. (1986). The analyses of natural microbial populations by ribosomal RNA sequences. *Adv. in Microb. Ecol.* **9,** 1–55.

Page, A. L., ed. (1982). "Methods of Soil Analysis. Part 2. Chemical and Microbiological Properties," 2nd ed., *Agronomy,* vol. 9. Am. Soc. Agron., Madison, Wisconsin.

Parkinson, D., Gray, T. R. G., and Williams, S. T. (1972). "Methods for Studying the Ecology of Soil Microorganisms." Blackwell, Oxford.

Rovira, A. D., and Campbell, R. (1974). Scanning electron microscopy of microorganisms on the roots of wheat. *Microb. Ecol.* **1,** 15–23.

Stanier, R. Y., Ingraham, J. L., Whealis, M. L., and Painter, P. R. (1986). "The Microbial World," 5th ed. Prentice-Hall, Englewood Cliffs, New Jersey.

Vance, E.D., Brookes, P.C., and Jenkinson, D.S. (1987). An extraction method for measuring soil microbial biomass C. *Soil Biol. Biochem.* **19,** 703–707.

Supplemental Reading

Burges, A., and Raw, F. (1967). "Soil Biology." Academic Press, London.

Burns, R. G., ed. (1978). "Soil Enzymes." Academic Press, London.

Burns, R. G., and Slater, H. (1982). "Experimental Microbial Ecology." Blackwell, Oxford.

Parkinson, D. (1982). Filamentous fungi. *Agron. Monogr.* **9,** 949.

Schmidt, E. L., and Paul, E. A. (1982). Microscopic methods for soil organisms. *Agron. Monogr.* **9,** 808.

Wollum, A. G. (1982). Cultural methods for soil organisms. *Agron. Monogr.* **9,** 782–802.

Chapter 4

Components of the Soil Biota

Introduction

The free-living components of the soil biota are the bacteria, fungi, algae, and the fauna. Additionally, there are the viruses; these grow only within the living cells of other organisms. Within each of the free-living components, there exists a broad range of morphological and physiological characteristics that has led to the naming of a large number of taxa for each group. As is true of any branch of science, the discipline of taxonomy is continually undergoing change, and consequently, taxonomic placements may differ rather sharply over time. In past years, the actinomycetes have often been given discussional status equivalent to that of the bacteria and fungi. Herein the actinomycetes are discussed as a subdivision of the bacteria. They are so treated in "Bergey's Manual of Determinative Bacteriology." For many years microbiologists have considered the blue-green "algae" a subdivision (cyanophyta) of the algae. Currently, most soil microbiologists (and "Bergey's Manual") designate them as cyanobacteria.

There are some 200 genera in the bacterial kingdom. No attempt will be made to discuss each genus separately; rather, emphasis will be given to genera common in soil, and to certain others that play key roles in the cycling of nutrients. The fungi are treated principally on morphological characteristics and at the class level. Herein the taxonomy of the soil fauna is treated very summarily. Emphasis is given to their feeding habits and to their roles as grazers and predators of the soil microbiota.

The remarks offered below are designed to give a broad overview of the soil biota, especially for those readers who have not had previous exposure to discussions of that biota.

Viruses

Viruses consist of RNA or DNA molecules within protein coats. Viral particles are metabolically inert and do not carry out respiratory or biosynthetic functions. They were first identified in the late nineteenth century by their ability to pass through filters capable of holding all known bacterial types, by their properties of a living organism, and by their ability to cause many plant and animal diseases. The crystallization of the tobacco mosaic virus in 1935 demonstrated that the properties of living cells can be found in agents that could be crystallized, like chemicals. They multiply only within host cells. They induce a living host cell to produce the necessary viral components; after assembly, the replicated viruses escape from the cell with the capability of attacking new cells. This ability to interact with host genetic material can make them very difficult to control. It is also the reason why they are so useful as genetic transfer agents in genetic engineering, as they can serve as transfer agents between a wide diversity of cells.

Viruses infect all categories of animals and plants, from humans to microbes. Those parasitizing bacterial cells commonly are termed bacteriophages, or simply, phages. Little is known about the field ecology of the viruses that infect soil organisms except that they can persist in soil as dormant units that retain parasitic capability. The ability of viral particles pathogenic to plants or animals to survive in soil and to move into the water table is of major concern to people. They also appear promising for use in the biological control of weeds and noxious insects. Once successful exploitation of viral parasites of undesirable hosts is achieved, it can be expected that efforts will be intensified to exploit the phages of bacterial species responsible for specific steps in soil processes such as nitrification.

Bacteria

Bacteria are the most numerous of the microorganisms in soil. Indeed, they are the most common of all the living organisms on the face of the earth. Bacteria lack nuclear membranes and are termed prokaryotic. Their nucleoplasm is not separated from the cytoplasm, as in fungi, protozoa, and other eukaryotes. Bacterial cell walls are composed principally of peptidoglycans, and reproduction is by binary fission. Genetic exchange is accomplished by conjugation and transduction. Conjugation involves the transfer of large portions of genetic material between donor and recipient cells in mating pairs. Transduction involves direct genetic transfer of DNA by viruses attacking bacteria (bacteriophages).

Both energy source and carbon source are useful for describing basic physiological differences among bacteria as well as among organisms generally. Those using light as their energy source are termed phototrophic; those deriving their energy from a chemical source, chemotrophic. If CO_2 is used as the cell carbon source, the organism is termed lithotrophic. If cell carbon is derived principally from an organic substrate, the organism is organotrophic. Essentially the same differentiation given in the terms lithotrophic and organotrophic is provided by the terms autotrophic and heterotrophic, respectively. The majority of known bacterial species are chemoorganotrophic and are commonly referred to as heterotrophs. Photolithotrophs include the higher plants, most algae, cyanobacteria, and green sulfur bacteria. Chemolithotrophs use diverse energy sources, e.g., NH_4^+, NO_2^-, Fe^{2+}, S^{2-}, and $S_2O_3^{2-}$. The obligate chemolithotrophs use the same basic physiological pathway (the Calvin cycle) found in most other organisms in metabolizing their own cell constituents. Their apparent inability to use any known external source of organic carbon is possibly linked to lack of permeases to move organic molecules across cell membranes. Organic molecules must be manufactured within the cell.

Physiological groupings for bacteria other than those based on energy and cell carbon sources are found in the literature. One of the best known is the autochthonous and zymogenous separation proposed by Winogradsky. Autochthonous organisms are defined as those growing slowly in soil containing no easily oxidizable substrate. Zymogenous species are those showing bursts of activity when fresh residues are added to soil. The concept is of some value, but to give individual genera or species autochthonous or zymogenous designation is difficult and even inadvisable. The literature abounds with instances in which a given species is assigned one status by some workers and the opposite status by others.

Ecological theory that examines the density of a species with respect to its food supply involves the concept of r and K selection. A species adapted to living under conditions of a bountiful energy source is designated as K selected. It would be exposed to selection pressures different from those affecting an r-selected organism living in uncrowded but possibly physically restrictive environments. The r-selected organisms exposed to flushes of substrate in an otherwise uncrowded environment would place a premium on high growth increase per unit of food rather than on competitiveness, as would be the case of a K-selected organism. Most soil organisms growing on flushes of substrate are usually r selected, whereas rhizosphere organisms and plant pathogens are usually K selected.

The morphological and staining properties of bacterial cells are useful taxonomically, but unfortunately by themselves they often fail to differentiate species that are profoundly different physiologically. Cell shape, whether rodlike, coccoid, or spirilliform, and presence or absence of flagellation and endospore formation are widely used descriptively. Tinctorially, response to the Gram stain is also useful. A designation G^+ is given to cells that retain crystal violet following subsequent exposures to KI solution, alcohol, and counterstain, and G^- for cells from which the crystal violet readily destains.

The following paragraphs offer a brief discussion of some of the better known bacterial genera. The reader is cautioned that the remarks offered concerning dominance and participation in soil processes are based on the published literature, which may be biased by ease of isolation of bacterial species or ease of their culture *in vitro*.

Arthrobacters

Members of the genus *Arthobacter* are the numerically predominant bacteria in soil, as determined by the plate count. Some estimates place their number as high as 40% of the total plate-count population. They are characterized by pleomorphism and Gram variability. Their cells are slender Gram-negative (G^-) rods in the early growth stage; later, the cells become very short Gram-positive (G^+) rods and G^+ coccoids. Very long rods and branching may be observed within the first 2 hr following inoculation on fresh substrates (Fig. 4.1). Many arthrobacters show a feeble motility that is often overlooked. As might be expected from their high numbers in soil, a wide variety of substrates are used in their oxidative metabolism. In one study, 85 of 180 compounds tested were utilized. The arthrobacters are slow growing (they form small colonies on agar surfaces) and are poor competitors in the early stages of residue decomposition, during which the easily decomposable materials (sugars, amino acids) are rapidly attacked by other genera.

Streptomycetes

Three genera, *Streptomyces, Pseudomonas,* and *Bacillus,* vie for the runner-up position to *Arthobacter* as commonly occurring bacteria in soil. Any one of the three may at various times account for 5 to 20% of the total bacterial count as determined by plate counting. Within the bacterial kingdom, the genus *Streptomyces* falls within the order Actinomycetales. The main features of five genera common in soil are shown in Table 4.1. About 90% of the actinomycete isolations from soil can be assigned to the genus *Streptomyces*. Its members produce a well-developed, compact, branched mycelium and compact colonies on agar plates. Reproduction is by copious production of aerial spores and by mycelial fragmentation.

Figure 4.1. Pleomorphism in *Arthrobacter globiformis*. (top) Very young cells at 2 hr. (bottom) Same culture at 30 hr.

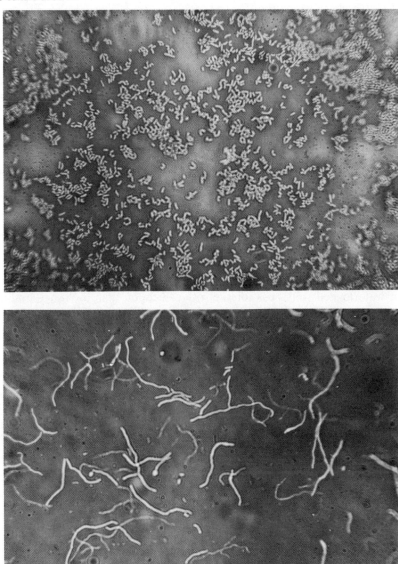

Table 4.1
Major Features of Some Genera of Actinomycetes Found in Soil[a]

Genus	Main features
Micromonospora	Filaments do not grow above medium; single spores produced in and on surface of medium; colonies rather slow growing on most media
Nocardia	Filaments unstable, fragmenting into bacteria-like units; filaments usually not growing above medium and spores rarely produced
Streptomyces	Long chains of spores formed on filaments growing above the medium; species very numerous in soil and many produce antibiotics
Streptosporangium	Spores formed in sporangia or in chains on the filaments above the medium; colony appearance similar to *Streptomyces*
Thermoactinomyces	Single spores formed on filaments above and within the medium; spores heat resistant; all species thermophilic

[a]From Gray and Williams (1971).

Streptomycetes are G^+ and are oxidative organotrophs. They are intolerant of waterlogged soils, less tolerant of desiccation than the fungi, and generally intolerant to acidity. Thus, the causal organism of potato scab, *S. scabies,* is often controlled in poorly buffered soils, such as sands, by sulfur or ammonium sulfate soil amendment, which results in a lowered soil pH.

Many streptomycetes produce antibiotics, variously antibacterial, antifungal, antialgal, antiviral, antiprotozoal, or antitumor. The soil microbiologist S. A. Waksman was awarded a Nobel Prize in medicine in 1942 for discovery of streptomycin. Streptomycetes produce dozens of differently named antibiotics and also geosmin. The last is responsible for the smell of freshly plowed soil and at least partly responsible for the musty smell of earthen cellars and old straw piles.

Pseudomonads

Members of the genus *Pseudomonas* are G^- straight or curved rods with polar flagellation. They are aerobic, except for denitrifying species that use nitrate as an alternate electron acceptor. Most species are organotrophs; a few are facultative lithotrophs, using H_2 or CO as an energy source. The pseudomonads occur not only in soil but also in fresh and marine waters. Some species cause plant disease; many nonpathogens are closely associated with plants. Pseudomonads, as a group, attack a wide variety of organic substrates, including sugars and amino acids, alcohols and aldose sugars, hydrocarbons, oils, humic acids, and many of the synthetic pesticides. Many species produce diffusible fluorescent pigments, especially in iron-deficient media. The fluorescein-linked siderophores have great affinity for Fe^{3+} and act similarly to ethylenediaminetetraacetic acid (EDTA) in iron transport. There is evidence that siderophore-producing pseudomonads can be used in the biological control of soilborne plant pathogens. In alkaline soils of low iron availability, the pseudomonad may deprive the pathogen of its iron requirement.

The closely related genus *Xanthomonas* embraces similar species, excepting that molecular oxygen is the only electron acceptor and nitrates are not reduced. *Xanthomonas* spp. are pathogenic to many plants. Within the genus there are a great many nomenspecies, distinguishable from the type species only by specificity of host plant reactions.

Sporulating Bacilli

Members of the genus *Bacillus* are G^+ to G-variable rods; most species are motile. Heat-resistant endospores are produced, and sporulation is not repressed by exposure to air. The sporulating bacilli are mostly vigorous organotrophs, and their metabolism is strictly respiratory, strictly fermentative, or both. There is great diversity within the genus, as shown by the array of products formed by different species during the course of glucose fermentation; the following are produced: glycerol, 2,3-butanediol, ethanol, H_2, acetone, and formic, acetic, lactic, and succinic acids. Some species have been described as facultative lithotrophs that use H_2 as an energy source in the absence of organic carbon. One species *(B. polymyxa)* has been found to fix dinitrogen (N_2). Several species produce lytic enzymes and antibiotics of the polypeptide class that are destructive of other bacteria. The toxin produced by *B. thuringiensis* is pathogenic to some insect larvae and is widely used as a biological control agent. *Bacillus macerans* has been used extensively in the retting of flax. *Bacillus anthracis* is a highly virulent animal pathogen. Temperature tolerance in the genus ranges from about -5 to $75°C$, tolerance to acidity from pH 2 to 8, and salt tolerance, to as high as 25% NaCl.

Clostridium is also a sporogenic genus. Most species are strict anaerobes, but a few are microaerophilic, forming small colonies on agar surfaces but not sporulating in the presence of air. The genus is of economic importance; its species are used commercially for the production of alcohols and commercial solvents. Several species, such as *C. butyricum* and *C. pasteurianum,* are known to fix N_2. The genus is widely distributed in soils, marine and freshwater sediments, manures, and also in animal intestinal tracts. Two well-known animal pathogens are *C. tetani* and *C. botulinum,* the spores of which can persist in soil for extended periods.

The above genera have been briefly described because of their very common occurrence in soil. They participate to varying degrees in transformations of carbon and other nutrients in soil. Inasmuch as there are hundreds of bacterial genera and thousands of species in soil, it is impractical to attempt any comprehensive coverage of the many taxa. The following genera are briefly noted to introduce some names that are commonly encountered in the soil microbiological literature:

Azotobacter is an aerobic organotroph capable of fixing N_2 asymbiotically. Other genera fixing N_2 asymbiotically but not as widely known are *Azomonas, Beijerinckia, Derxia,* and *Azospirillum. Rhizobium* and *Bradyrhizobium* are widely known genera that fix N_2 symbiotically. Their cells invade root hairs of leguminous plants, induce nodule formation, and grow as intracellular symbionts. A related genus, *Agrobacterium,* induces galls or other hypertrophies, such as very hairy roots, on plants but does not fix N_2. *Nitrosomonas* and *Nitrobacter* are chemolithotrophic genera long known to cause nitrification in soil. The former converts NH_4^+ to NO_2^-, and the latter, NO_2^- to NO_3^-.

Lactobacillus is a fermentative organotroph that merits mention because of its common association with plant herbage. Its lactic acid production is variously exploited in making silages, cultured buttermilks, sauerkrauts, and sourdoughs. *Enterobacter,* also fermentative, is commonly found in animal feces and sewage, but some species are widely distributed in soil and on plants. Profuse growth of *E. cloacae* on sugar cane cuttings and in frostbitten cotton bolls can produce plant material yielding a dust toxic to cane-mill and cotton-gin workers. *Cytophaga* is a member of the gliding bacteria. Some species can glide as rapidly as 10 μm min^{-1} on moist surfaces, such as on an agar plate. The genus is active in cellulose decomposition. Various bacterial genera active in sulfur and metal transformations are named in later chapters.

Cyanobacteria

The kingdom Procaryotae embraces two divisions, the bacteria and the cyanobacteria (Buchanan and Gibbons, 1974). Genera named in preceding paragraphs belong to the bacteria. The cyanobacteria are photosynthetic prokaryotes containing chlorophyll, and also phycobiliprotein pigments such as phycocyanin. Algologists contend that the blue-greens should be designated as algae inasmuch as they contain chlorophylls that are similar to those found in higher plants and algae. The photosynthetic bacteria contain a range of bacteriochlorophylls and do not liberate free O_2 during photosynthesis, as do the blue-greens and higher plants.

The cyanobacteria exist in unicellular, colonial and filamentous forms (Fig. 4.2). Single cells, reproductive cells, or units and filamentous forms not enclosed in rigid sheaths often show a gliding motility. The cellular cytoplasm is structured, with paired photosynthetic lamellae (termed thylakoids) whose outer surfaces bear characteristic pigment-containing granules. Cell walls are complex or multilayered and contain peptidoglycan and a mucopolymer that is dissolved by lysozyme, as are the cell walls of G^- bacteria. Outer mucilaginous sheaths, sometimes without sharply defined outer limits, commonly occur; they consist of mucopolysaccharides and pectic acids. Reproduction is by cell division and involves several kinds of specialized cells: endospores, exospores, and akinetes. Unicellular cyanobacteria variously show binary fission, multiple fission, or serial release of apical cells as exospores. Filamentous forms reproduce by random fragmentation and by the terminal release of short, motile chains of cells termed hormogonia. Certain of the filamentous forms produce akinetes and heterocysts. Akinetes are oversized resting vegetative cells that following germination produce a hormogonium. Heterocysts are nonreproductive cells differing from adjoining vegetative cells by the possession of refractile granules and a thicker outer wall.

Bold and Wynne (1979) have classified the blue-greens into three orders: Chamaesiphonales (sporogenous, producing endospores or exospores), Chroococcales (unicellular or colonial, and asporogenous), and Oscillatoriales (filamentous and asporogenous). Other writers have used filament

Figure 4.2. Vegetative cells (V), heterocysts (H), and spores (S) of the N_2-fixing *Anaebaena cylindrica*. (From Stewart, 1977.)

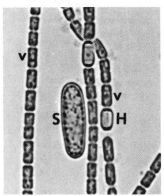

characteristics to subdivide the last-named into Oscillatoriales and Stigonematales. The cyanobacteria are ubiquitous in their distribution, occurring in saline and fresh waters, in soil, and on bare rocks and sand. On soil parent materials they are important as primary colonizers, either alone or as symbionts of fungi in lichens. Cyanobacteria also occur within the plant bodies of certain liverworts, water ferns, and angiosperms. In some ecosystems, cyanobacteria are of great significance because of their ability to fix N_2. Although certain nonheterocystous genera such as *Gloeothece* are known to fix N_2, there is good evidence that heterocysts when present are especially active sites for fixation and that fixed nitrogen may be transferred from heterocysts to adjacent vegetative cells.

A discussion of biological nitrogen fixation by cyanobacteria and other organisms is given in Chapter 10.

Fungi

The fungi embrace eukaryotic organisms variously referred to as molds, mildews, rusts, smuts, yeasts, mushrooms, and puffballs. Of the soil organisms, the fungi as a group are the organotrophs primarily responsible for the decomposition of organic residues, even though in plate counts they are outnumbered by the bacteria. The fungi are divided into subgroups by morphological characteristics. Typically, the fungi form slender filaments or hyphae, either septate or nonseptate, and commonly multinucleate. Collectively, the hyphae constitute the mycelium, soma, or thallus. Such bodies may reach several decimeters in diameter and are easily visible. Fungal cell wall constituents are listed in Table 4.2. Features characterizing eight fungal classes commonly present in soil are summarized in Table 4.3. It must be emphasized that the following remarks offer only a very limited treatment of the soil fungi. For more adequate coverage, the reader is referred to Ainsworth and Sussman (1965–1973), Alexopoulos and Mims (1979), Ross (1979), Gilman (1957), and Domsch *et al*. (1981).

Slime Molds

The Acrasiomycetes are the cellular slime molds. The unit of structure is an uninucleate amoeba that feeds by engulfing bacteria. Single cells characteristically aggregate into a pseudoplasmodium in which the cells do not fuse but behave as a mobile communal unit. The pseudoplasmodium changes into a fruiting structure, the sporocarp, that bears asexual spores. Acrasiomycetes are found on decaying plant materials in moist environments. Inasmuch as they feed on the bacteria, they cannot be considered

Table 4.2
Components of the Fungal Cell Wall

Cell-wall category	Taxonomic group	Representative genera
Cellulose–Glycogen	Acrasiomycetes	*Dictyostelium*
Celluose–β-glucan	Oomycetes	*Phytophthora, Pythium, Saprolegnia*
Cellulose–chitin	Hyphochytridiomycetes	*Rhizidiomyces*
Chitin–chitosan	Zygomycetes	*Mucor, Phycomyces, Zygorhynchus*
Chitin–β-glucan	Chytridiomycetes	*Allomyces, Blastocladiella*
	Ascomycetes	*Neurospora, Alllomyces*
	Deuteromycetes	*Aspergillus*
Mannan–β-glucan	Ascomycetes	*Schizophyllum, Fomes, Polyporus*
Chitin–Mannan	Basidiomycetes	*Saccharomyces, Candida*
Galactosamine–galactose polymers	Trichomycetes	*Sporobolomyces, Rhodotorula, Amoebidium*

primary agents of organic matter decomposition. There are eight genera. *Dictylostelium* is relatively easy to isolate from soil and is perhaps the best known.

The Myxomycetes are the true slime molds. They form an acellular creeping plasmodium that suggests that they should be assigned to the animal kingdom. They are animal-like in their feeding, plasmodial form but fungus-like in their reproductive structures and spore formation. The group is widely distributed in soil, especially in association with decaying vegetation in cool, moist sites. Some species develop on herbage, and others on animal dung. Plasmodia attaining diameters of 50 cm or more sometimes occur as unsightly slimy globs on grass. The genus *Physarum*, containing more than 100 species, is the largest genus. It has been used extensively in laboratory studies on fungal morphology and physiology.

Flagellate Fungi

Oomycetes are to be found in water and soil, and many are highly destructive plant pathogens. They are unique in that they produce biflagellate asexual motile spores, termed zoospores. *Pythium* and *Phytophthora* are genera commonly encountered. *Pythium debaryanum* causes damping-off of seedlings; *Phytophthora infestans,* potato blight; and *Plasmopara viticola,* grape mildew. The last-named was the first organism controlled by a fungicide, namely, Bordeaux mixture, a copper arsenate compound. The genus *Saprolegnia* belongs to a subgroup commonly called the water molds, but many of its species also occur in soil.

Table 4.3
Characteristics of Fungi Associated with Soils and Plants

Group	Class	Features	Representative soil genera
Slime molds	Acrasiomycetes	Uninucleate myxomoeba, pseudoplasmodium sporocarp bears asexual spores; found as bacterial feeders on decaying vegetation	*Dictylostelium*
	Myxomycetes	True slime molds; acellular creeping plasmodium, which is animal-like but produces fungal-like reproductive structures	*Physarum*
Flagellate fungi	Oomycetes	Produce biflagellate oospores within a sporangium; reproduction by game tangy	*Pythium, Plasmopara, Phytophthora, Saprolegnia*
	Chytridiomycetes	Aquatic habitats; polarly uniflagel late motile zoospores; some parasitic.	*Allomyces, Rhizophydium*
Sugar fungi	Zygomycetes	Nonseptate coenocytic hypha; internal sporangiospores and thick-walled zygospores; mostly saprobic; some phytophathogenic or parasitic on other fungi	*Mucor, Mycotypha, Rhizopus, Zygorhyncus*
Higher fungi	Ascomycetes	Sac or ascus containing ascospores formed from karyogamy and meiosis, e.g., sexual stages, unicellular yeasts, or septatic hyphae	*Endothia, Ceratocystus, Claviceps, Saccharomyces*

Table 4.3 (Continued)

Group	Class	Features	Representative soil genera
	Basidomycetes	Septate hyphae, sexual spores, produced by meiosis, held externally on basidium; many mushrooms, rusts, etc.	*Agaricus, Poria, Boletus, Fomes*
Fungi Imperfecti	Deuteromycetes	Septate hyphae reproduce only by asexual conida; if sexual stage found, usually ascomycete	*Aspergillus, Trichoderma, Penicillium, Helmimthosporium, Fusarium, Arthrobotrys*
	(Mycelia Sterilia)	Subgroups of Deuteromycetes; no conidia produced; reproduce by hyphal fragmentation	*Rhizoctonia*

The Chytridiomycetes are especially prevalent in aquatic habitats but also commonly occur in soil. Their production of polarly uniflagellate motile zoospores distinguishes them from other fungi. Some members are parasitic on algae, higher plants, or insect larvae. Among the genera commonly found in soil are *Allomyces* and *Rhizophydium*.

Sugar Fungi

The Zygomycetes are called the sugar fungi because of their fermentations of diverse carbohydrate substrates. They usually produce a well-developed mycelium of coenocytic hyphae, asexual sporangiospores, and thick-walled resting zygospores. The group is mostly saprobic, but some are phytopathogenic, some parasitic on other fungi, and some produce animal-trapping mechanisms. The Mucorales (the largest order of the Zygomycetes) are important economically; individual species are used for the commercial production of alcohols and organic acids, such as lactic, citric, oxalic, or fumaric. Species of several genera (e.g., *Mucor, Mycotypha*) remain in the yeast form when grown anaerobically or even when grown aerobically in the presence of high concentrations of glucose (>5%). *Rhizopus nigricans* is the common bread mold.

Higher Fungi

The Ascomycetes and the Basidiomycetes are sometimes called the "higher fungi." The former are distinguished from other fungi by the formation of a sac or ascus within which ascospores (commonly four or eight) are formed following sexual reproduction. Included in the Ascomycetes are the ascus-forming yeasts, morels and truffles, cup fungi, powdery mildews, and many of the common dark-colored molds. Many species are saprophytic in soil, with the mycelium remaining underground but at intervals sending up very large fruiting bodies. The Ascomycetes have a wide range of impacts on people. Many are parasitic on plants, causing root rots, corn ear rots, brown rots of stone fruits, and powdery mildews. The ascomycete *Endothia parasitica* is destructive of chestnut trees, and *Ceratocystis ulmi*, of the American elm. The ergot fungus *(Claviceps purpurea)* invades fruiting structures of grasses and produces alkaloids toxic to humans and other animals. Species of *Chaetomium* are especially destructive of cellulose fabrics. On the beneficial side, the fermenting activities of the yeasts *(Saccharomyces)* have long been exploited in the beer and wine industries and in bakeries.

The Basidiomycetes include a wide assortment of fungi, notably mushrooms, puffballs, stinkhorns, shelf and bracket fungi, bird's-nest fungi, jelly fungi, smuts, and rusts. The Basidiomycetes differ from other fungi by their production of spores (basidiospores) externally on a specialized structure called the basidium. As plant parasites, the Basidiomycetes cause astronomical losses of crop plants and trees. Particularly damaging are the smuts and stem rusts of cereals, diseases of forest and shade trees, and the rotting of lumber and of wooden structures. Species in the order Uredinales are especially villainous as causal agents of plant rusts. Thousands of species are involved, among them the black stem rust of cereals and the white pine blister rust. On the beneficial side are their mycorrhizal-forming relationships with plants, and their commercial exploitation wherein mushrooms are grown for edible food. *Agaricus brunnescens* is the species widely grown commercially in horse manure compost. Among many others, *Boletus edulis* is a commonly collected edible "wild" species. There are many poisonous genera (e.g., *Amanita, Coprinus, Clitocybe, Psilocybe*) in the Agaricales. This order also includes many mycorrhizal fungi that are discussed in Chapter 11.

The Basidiomycetes are extremely vigorous decomposers of woody material, including standing and felled timber, tree stumps, and lumber products. Fungi decomposing cellulose but not lignin are causal agents of wood brown rots, so named because the partly oxidized dark-colored lignin remains. White rots are caused by fungi destroying both cellulose and lignin. Among genera important in wood rotting are the following: *Fomes, Polysporus, Poria,* and *Armillariella.*

Imperfect Fungi

The Deuteromycetes embrace fungi that have septate hyphae but thus far are known to reproduce only by means of conidia. Because of their lack of a sexual phase, they are called "Fungi Imperfecti." A sexual stage may be discovered long after the initial naming of a deuteromycete and it then becomes necessary to assign the species to another class, usually to the Ascomycetes, less often to the Basidiomycetes. Some mycologists believe that many of the imperfect fungi represent conidial stages of ascomycetes whose sexual stages are rarely formed or observed in nature or else have been dropped from the life cycle altogether. Inasmuch as the parasexual cycle has been discovered in some of the imperfects, it is also possible that such fungi have never had a sexual stage. A subgroup included in the Deuteromycetes is designated the "Mycelia Sterilia." Its members fail to produce conidia. Reproduction is by fragmentation of hyphae.

The majority of the Deuteromycetes are saprobic in soil, but again, many are parasitic on other fungi, higher plants, and humans and other animals. As a group the imperfect fungi are too numerous and too diverse to lend themselves to discussion within a few paragraphs. The following chapters will cite some genera participating in specific events or transformations. Generic names commonly encountered in the soil microbiological literature include *Aspergillus, Penicillium, Trichoderma, Fusarium, Cladosporium, Arthrobotrys, Gliocladium,* and *Helminthosporium.* The genus *Rhizoctonia* exemplifies the Mycelia Sterilia subgroup. A summary of the major classification criteria is given in Table 4.3, and representative structures are shown in Fig. 4.3.

Algae

The eukaryotic algae have been given names such as pond scum, water moss, seaweed, and red tide. Algae are the simplest of the chlorophyllus eukaryotes, distinguishable from all other green plants by sexual characteristics. In unicellular algae, the entire organism may function as a gamete (sexually reproductive cell). In multicellular algae, gametes are produced in either unicellular or multicellular gametangia, with every gametangial cell producing a gamete. In asexual reproduction, algae produce flagellated and/or nonmotile spores.

Figure 4.3.   Fungi commonly found on plate-culture isolates of soil. (From Gilman, 1957.)

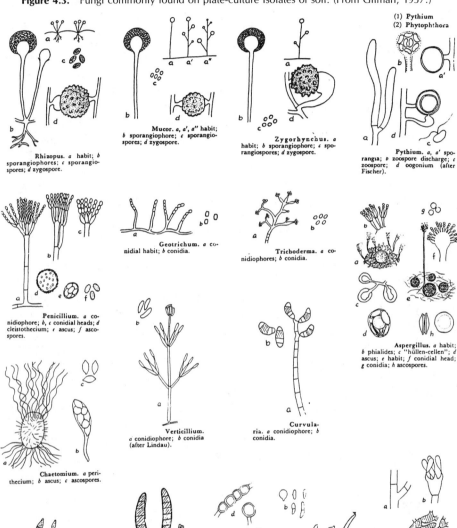

(1) **Pythium**
(2) **Phytophthora**

Rhizopus. *a* habit; *b* sporangiophores; *c* sporangiospores; *d* zygospore.

Mucor. *a, a', a''* habit; *b* sporangiophore; *c* sporangiospores; *d* zygospore.

Zygorhynchus. *a* habit; *b* sporangiophore; *c* sporangiospores; *d* zygospore.

Pythium. *a, a'* sporangia; *b* zoospore discharge; *c* zoospore; *d* oogonium (after Fischer).

Penicillium. *a* conidiophore; *b, c* conidial heads; *d* cleistothecium; *e* ascus; *f* ascospores.

Geotrichum. *a* conidial habit; *b* conidia.

Trichoderma. *a* conidiophores; *b* conidia.

Aspergillus. *a* habit; *b* phialides; *c* "hüllen-cellen"; *d* ascus; *e* habit; *f* conidial head; *g* conidia; *h* ascospores.

Chaetomium. *a* perithecium; *b* ascus; *c* ascospores.

Verticillium. *a* conidiophore; *b* conidia (after Lindau).

Curvularia. *a* conidiophore; *b* conidia.

Alternaria. Conidiophores and conidia.

Helminthosporium. Conidiophores and conidia.

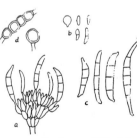

Fusarium. *a* conidial head; *b* microconidia; *c* macroconidia; *d* chlamydospores.

Rhizoctonia. *a* hypha; *b* basidium and spores; *c* sclerotial hyphae.

The algae are unquestionably the most widely distributed of all green plants. They are predominantly aquatic, being found in fresh, brackish, and salt waters. Salt tolerance ranges from salt concentrations near zero to near 100%. Aquatic algae may be either planktonic (free-floating) or benthic (attached to the bottom). Terrestrial forms occur on and in the soil, on rocks, mud, and sand, on snowfields and buildings, and attached to plants (stems, bark, leaves) and animals. Subsurface soil samples kept moist and under illumination commonly develop algal blooms. Most algal units that are found below ground are dormant forms, but some are known to be facultative organotrophs.

Early classification relied heavily on pigmentation; subgroupings were designated as green, brown, and red algae, and the diatoms. The last named are characterized by a rigid cell wall composed of silica with an organic coating. Classification of the diatoms (Bacillariophyceae) is based on the composition of the cell wall and the patterns of the markings thereon. Current classifications for the algae rely on pigmentation, cell wall constituents, cellular organization, storage products, and flagellation. Bold and Wynne (1979) have recognized 10 major subdivisions of the eukaryotic algae; they have prepared an excellent summary of divisional characteristics (Table 4.4). There are many generalized discussions, among them those by Tiffany (1958), Prescott (1968), Round (1973), and Stewart (1977), of algal characteristics, physiology, and taxonomy.

Lichens

A lichen is a symbiotic association of a fungus and an alga with the two so intergrown as to form a single thallus, or undifferentiated body. The fungal member is usually an ascomycete, less often a basidiomycete or a deuteromycete. It has been estimated that there are roughly 18,000 species of lichenized ascomycetes but only about 15,000 species that are nonlichenized. The photosynthetic partner (the photobiont) may be either a eukaryote or a prokaryote. In most lichens, it is a green alga and usually of the genus *Trelouxia;* less often it is a cyanobacterium, usually of the genus *Nostoc*. A lichen association is given taxonomic status and binomial designation. Major subgroupings are made on the basis of the fungal symbiont; thus there are ascolichens, basidiolichens, and deuterolichens.

The photobiont in the symbiosis captures energy, and if it is a cyanobacterium, it may also fix N_2. In return for carbon and at times for nitrogen, the fungal partner (the mycobiont) is believed to furnish mineral nutrients and to help regulate the water and light requirements of its symbiont. The

Table 4.4
Some Algal Divisions and Their More Significant Characteristics[a]

Division	Common name	Pigments, plastid organizations in photosynthetic species	Cell wall	Habitat[b]
Chlorophycophyta	Green algae	Chlorophyll *a*, *b*; α- β-, and γ-carotenes, several xanthophylls; 2–5 thylakoids/stack	Cellulose in many [β- (1→4) glucopyranoside], hydroxyproline glycosides, xylans and mannans, or wall absent; calcified in some	FW, BW, SW t.
Charophyta	Stoneworts	Chlorophyll *a*, *b*; α-, β-, and γ-carotenes, several xanthophylls; thylakoids variably associated	Cellulose [β (1→4)-glucopyranoside]; some calcified	FW, BW
Euglenophycophyta	Euglenoids	Chlorophyll *a*, *b*; β-carotene, several xanthophylls; 2–6 thylakoids/stack, sometimes many	Absent	FW, BW, SW, T
Phaeophycophyta	Brown algae	Chlorophyll *a*, *c*; β-carotene, fucoxanthin, and several other xanthophylls; 2–6 thylakoids/stack	Cellulose, alginic acid, and sulfated mucopolysaccharides (fucoidan)	FW (very rare), BW, SW

66

Chrysophycophyta	Golden and yellow-green algae (including diatoms)	Chlorophyll a, c (c lacking in some); α-, β-, and ε-carotene, several xanthophylls, including fucoxanthin in Chrysophyceae, Bacillariophyceae, and Prymnesiophyceae; 3 thylakoids/stack	Cellulose, silica, calcium carbonate, mucilaginous substances, and some chitin, or wall absent	FW, BW, SW, T
Pyrrhophycophyta	Dinoflagellates	Chlorophyll a, c; β-carotene, several xanthophylls; 3 thylakoids/stack	Cellulose or absent; mucilaginous substances	FW, BW, SW
Cryptophycophyta	Cryptomonads	Chlorophyll a, c; α-, β- and ε-carotene, distinctive xanthophylls (alloxanthin, crocoxanthin, monadoxanthin), phycobilins; 2 thylakoids/stack	Absent	FW, BW, SW
Rhodophycophyta	Red algae	Chlorphyll a, (d in some Florideophycidae); R- and C-phycocyanin, allophycocyanin; R- and B-phycoerythrin; α- and β-carotene, several xanthophylls; thylakoids single, not associated	Cellulose, xylans, several sulfated polysaccharides (galactans); calcification in some	FW (some), BW, SW (most)

[a]Adapted from Bold and Wynne (1979).

[b]BW, brackish water; FW, fresh water; SW, salt water; T, terrestrial.

association permits the survival of the two symbionts in harsh environments in which neither could survive separately.

Lichens are found worldwide on inhospitable rock, soil, and other surfaces, such as fence posts, roof tops, foliage, and tree barks. In the Arctic, they constitute at least part of the winter forage for reindeer. The lichens are commonly among the first colonizers of bare rocks and soil parent materials and thus are important in the early stages of pedogensis. In general, lichens are not ascribed major roles in the cycling of carbon and minerals in highly productive ecosystems. Ahmadjian and Hale (1973) have offered an in-depth discussion of the morphology, physiology, and taxonomy of the lichens.

Protozoa

Protozoa are unicellular animals, most of which are microscopic in size but some attaining macroscopic dimensions. The group is greatly diverse in morphology and feeding habits but does show commonality in that all require water envelopment for metabolic activity. Some 30,000 living and as many fossil species have been described. Five main subgroups are commonly recognized: flagellates, ciliates, naked and testate amoebae, and sporozoa. The last is wholly parasitic. A resume (Table 4.5) of the protozoa occurring in soil or in association with other soil organisms has been prepared by Stout *et al.* (1982). It may be noted that the slime molds are included in their resume; these molds have been discussed above as fungi. Another controversial group are the photosynthetic euglenoids. *Euglena gracilis* and related species are claimed both by protozoologists (e.g., Nisbet, 1984) and algologists (e.g., Bold and Wynne, 1979). We have chosen to treat the euglenoids as algae.

Free-living protozoa in soil feed on dissolved organic substances and on other organisms. Many feed wholly by grazing and predation. The soil ciliates depend primarily on bacteria as food; some feed additionally on yeasts and other protozoa, and even on small metazoa such as rotifers. Amoebae feed on bacteria, other protozoa, yeasts, fungal spores, and algae. In laboratory studies, a single protozoan has been observed to consume several hundred to several hundred thousand bacterial cells per hour. The number consumed is a function both of protozoan and bacterial body size.

The soil protozoa have an effect on the structure and the functioning of microbial communities. The rise in bacterial numbers commonly observed following addition of fresh residues to soil is almost always followed

Table 4.5
Principal Free-Living Soil Protozoan Groups[a]

Small flagellates
  Ubiquitous, e.g., *Oikomonas* (one flagellum), *Bodo* (two flagella, one trailing)
Naked amoebae
  Ubiquitous
    Small monopodal: *Vahlkampfia, Naegleria, Hartmannella* (including parasitic species)
    Multipodal: *Nuclearia*
    Pellicle layer: *Thecamoeba*
    Slime molds: *Dictyostelium*

Ciliates
  Reflect soil structure, moisture, and aeration
    Trichostomes: almost ubiquitous, e.g., *Colpoda* (oval, flat body with indented cytostome), *Leptopharynx* (small, flat, with coarse ciclia in furrows)
    Small edaphic ciliates
      Chilodenella, humped, with cilia restricted to ventral surface
      Cyrtolophosis, ovoid, with cytostome in depression in anterior end
    Larger and more complex hypotrichs: *Uroleptus, Keronopsis, Gonostomum*
    Sessile peritrichs, e.g., *Vorticella striata*
    Predatory species: *Spathidium, Bresslaua*
    Histophage: *Tetrahymena rostrata* (facultative parasite)
    Suctoria, sessile adult with ciliated larva, e.g., *Podophrya*

Testacea
  Shape of the test commonly reflects the soil moisture regime
    Globose or hemispherical tests, e.g., *Cyclopyxis, Phryganella*
    Oval, with subterminal aperture, e.g., *Trinema, Corythion, Centropyxis*
    High-vaulted tests, normally associated with forest litters and mosses, e.g., *Nebela, Difflugia*
    Flattened, planoconvex (in profile) tests, normally in open structures, e.g., litter (*Arceila, Microchlamys*)

[a]From Stout *et al.* (1982).

by a rise in protozoan numbers. Selective feeding by protozoa may alter the mix of bacterial genera. In laboratory microcosms, decomposition of organic residues proceeds more rapidly in the presence of both protozoa and bacteria than with bacteria only. It is assumed that protozoan grazing keeps that population physiologically young, hence more efficient in residue decomposition. Animal ecologists have long espoused the hypothesis that predators have beneficial effects on their prey, collectively but not individually. Protozoans may accelerate nutrient cycling, and their active motility in the soil water may be helpful in providing bacteria with dissolved oxygen and nutrients.

Soil Metazoa

Soil microbiology includes the study not only of the microflora and protozoa, but also of many small soil-dwelling multicellular animals (the metazoans), such as nematodes, millipedes, centipedes, rotifers, mites, annelids, spiders, and insects (Fig. 4.4). The smallest members of this miscellany (e.g., some nematodes) are able to move through existing soil pores without disturbing the soil particles and are often designated, together with the soil protozoa, as the soil microfauna. Permanent soil dwellers causing at least some soil disturbances are commonly called the soil mesofauna.

The nematodes, also called roundworms, thread worms, or hair worms, are the most numerous of the soil metazoans. Their numbers may reach to several million per square meter. Most soil forms are less than 0.05 mm wide and 2 mm long. Many are parasitic on higher plants and animals. In soil, free-living forms are voracious feeders on both the microflora and other fauna. Earthworms often constitute the major portion of the invertebrate biomass in soil and when present are active in processing litter and distributing organic matter throughout the soil. Enchytraeids (0.2–0.8 mm wide and 5–15 mm long) are common in moist forest litter and soil. Mites, collembola, ants, and insect larvae also constitute important segments of the soil mesofauna.

The importance of the mesofauna in soil lies in its effects on the physical and chemical properties of soil and detritus and on the structures of microbial communities. Earthworms, ants, and termites fragment and transport organic matter into deeper soil layers. Millipedes, mites, isopods, and collembolans also participate in litter fragmentation. Metazoans that channel through the soil, or mix litter with soil, affect water intake and percolation rates. Those that fragment litter, especially woody litter as in tree stumps or logs, greatly increase litter surface-area exposure to rain, dew, and the soil microflora. The mesofauna, as the microfauna, can directly affect the structure of microbial communities. Bacterial intake rates by nematodes have been estimated to be as high as 5000 cells min^{-1}, and total consumption of the microflora by bacteriovores and fungivores as high as 50% of the estimated annual production of microfloral biomass.

Exclusion of the soil fauna from litter-enriched soil usually decreases the soil respiration rate, sometimes markedly, but the mineralization accomplished directly by the soil fauna is small compared to the microfloral activity. It has been estimated as only a few percent of the total. Mechanisms involved in the faunal acceleration of decomposition by indirect effects, such as by changing the physical and chemical properties of soil and litter and the composition of the soil microflora, are not yet fully

Figure 4.4. Numbers of animals per square meter in soil of a European grassland, plotted on a logarithmic scale. Spiders and centipedes are wholly carnivorous, and ants predominantly so; beetles, fly maggots, mites, and nematodes range widely in their diet according to species; the remaining groups feed largely on decaying organic matter, many collembola are closely associated with fungi and other microorganisms. (Adapted from Kevan, 1965.)

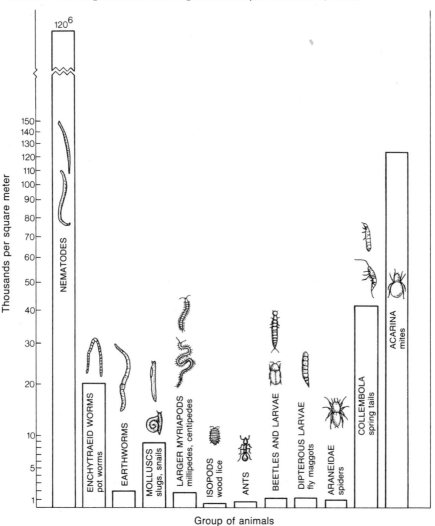

understood. Measurements obtained in closed systems cannot be extrapolated directly to the field. In complex systems such as soil, any of several factors may be involved. When several are acting simultaneously, any single factor can be expected to be less influential than indicated by measurements in closed systems.

References

Ahmadjian, V., and Hale, M. E., eds. (1973). "The Lichens." Academic Press, New York.
Ainsworth, G. C., and Sussman, A. S., eds. (1965–1973). "The Fungi: An Advanced Treatise" 4 vols. Academic Press, New York.
Alexopoulos, C. J., and Mims, C. W. (1979). "Introductory Mycology." Wiley, New York.
Bold, H. C., and Wynne, M. J. (1979). "Introduction to the Algae." Prentice-Hall, Englewood Cliffs, New Jersey.
Brock, T. D., Smith, D. W., and Madigan, M. T. (1984). "Biology of Microorganisms," 4th ed. Prentice-Hall, Englewood Cliffs, New Jersey.
Buchanan, R. E., and Gibbons, N. E., eds. (1974). "Bergey's Manual of Determinative Bacteriology," 8th ed. Williams & Wilkins, Baltimore, Maryland.
Domsch, K. H., Gams, W., and Anderson, T.-H. (1981). "Compendium of Soil Fungi." Academic Press, New York.
Gilman, J. C. (1957). "A Manual of Soil Fungi," 2nd ed. Iowa State Univ. Press, Ames.
Gooday, D. W. (1979). In "Microbial Ecology: A Conceptual Approach" (J. M. Lynch and N. J. Poole, eds.), pp. 5–21. Wiley, New York.
Gray, T. R. G., and Williams, S. T. (1971). "Soil Microorganisms." Oliver & Boyd, Edinburgh.
Kevan, D. K., McE. (1965). The soil fauna—its nature and biology. In "Ecology of Soil Borne Pathogens" (K. F. Baker and W. C. Snyder, eds.), pp. 33–51. Univ. of California Press, Berkeley.
Nisbet, B. (1984). "Nutritional Feeding Strategies in Protozoa." Croom Helm, London and Canberra.
Prescott, G. W. (1968). "The Algae: A Review." Houghton Mifflin, Boston, Massachusetts.
Ross, I. K. (1979). "Biology of the Fungi: Their Development, Regulation, and Associations." McGraw-Hill, New York.
Round, F. E. (1973). "The Biology of the Algae." St. Martin's Press, New York.
Stewart, W. D. P., ed. (1977). "Algal Physiology and Biochemistry." Univ. of California Press, Berkeley and Los Angeles.
Stout, J. D., Bamforth, S. S., and Lousier, J. D. (1982). Protozoa. Agron. Monogr. 9.
Tiffany, L. H. (1958). "Algae: The Grass of Many Waters." Thomas, Springfield, Illinois.
Winogradsky, S. N. (1949). Microbiologue du sol: problémes et méthodes. Massor et cie, Paris.

Supplemental Reading

Anderson, R. V., Coleman, D. C., and Cole, C. V. (1981). Effects of saprotrophic grazing on net mineralization. Ecol. Bull. 33, 201–216.

Andrews, J. H. (1984). Relevance of r- and k-theory to the ecology of plant pathogens. *In* "Current Perspectives in Microbial Ecology" (M. J. Klug and C. A. Reddy, eds.). Am. Soc. Microbiol., Washington, D.C.

Peterson, H. (1982). Quantitative ecology of microfungi and animals in soil and litter. *Oikos* **39,** 288–482.

Chapter 5

Occurrence and Distribution of Soil Organisms

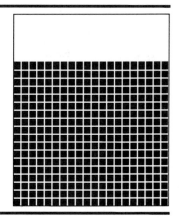

Introduction

Much time and effort have been spent in determining the kinds and numbers of organisms in soil and the sites colonized by them. Over geological time, microorganisms have had great opportunity to become distributed worldwide. Such geographical differences as now exist are reflective of long-term adaptations to the environment. Given similar site conditions, intraregionally as well as interregionally, similar soil biotas can be expected. Most members of the biota are organotrophic, and the availability of organic carbon largely determines both the numbers and the distributional patterns of organisms in soil. The major sources of carbon input are live plant roots and the organic residues contributed to soil during and following plant growth. Two very important processes linked to microbial associations with roots are those of nitrogen fixation and the formation of mycorrhizae; these two topics are given extended discussion in Chapters 10 and 11, respectively. In this chapter, attention is given to distributional and structural differences in the soil biota, with little or no attention to the nutrient transformations mediated by the organisms. These transformations are discussed in the chapters immediately following.

Geographical Differences in Soil Biota

Wind, water, and animal transport, in existence since before the advent of humans, foster worldwide distribution of microorganisms. Microbes are subject to airborne transport over long distances either independently, in association with dust particles, or in aerosols. Transocean transport of dust has been documented, and microorganisms are recoverable from snowflakes falling at high altitudes even after long intervals of precipitation. Ocean currents move debris between continents, and migratory birds traverse great distances. Thus almost any microbial species has opportunity for ubiquitous distribution. Most microbiologists need only to go to their own backyards to secure isolates of almost any microbial taxon. Nevertheless, there are some geographical disparities.

Among the fungi, *Penicillium* is more abundant in temperate and cold temperate climates than *Aspergillus,* whereas in warm regions the reverse is true. Russian workers have reported sporulating bacilli and actinomycetes as abundant in warm, dry soils and extremely scarce in cold, wet soils. In the United States, intraspecific differences in the optimum growth temperature of sporulating bacilli and nitrifying bacteria have been noted for isolates from northern and southern soils. Factors other than water and temperature are also influential. In the tropics, the *Fusarium* wilt of bananas is restricted to those soils containing the clay smectite. Cyanobacteria are commonly found in neutral to alkaline soils but only rarely in acidic soils. Such differences as do exist should not obscure the fact that for similar environments worldwide, essentially similar soil biotas are present.

Distribution of Organisms within the Soil Profile

Typically, microorganisms decrease with depth in the soil profile, as does organic matter. The bacterial numbers determined by direct and plate counts in a natural grassland are shown in Fig. 5.1. Comparison of the two curves shows that the plate count at the 5-cm depth is roughly 3% of the direct count, and at 75 cm, roughly 1%. The linear drop with depth in the direct count versus a curvilinear drop in the cultural (plate) count suggests that organisms at lower depths have a greater proportion either of dormant propagules or of species or variants incapable of growth on soil extract agar.

For soils generally, the population density does not continue to decrease to extinction with increasing depth, nor does it necessarily reach a constant or very slowly declining density. Fluctuations in density commonly occur at lower horizons. In alluvial soils, populations fluctuate with textural changes; organisms are more numerous in silty or silty clay horizons than in intervening sandy or coarse sandy horizons. In soil profiles above a perched water table, organisms are more numerous in the zone immediately above the water table than in higher zones. Given a water table a few meters below the soil surface and a plant such as mesquite, which

Figure 5.1.   Total bacterial counts for differing profile depths in a grassland soil. (Left) Bacterial numbers determined by plate count on soil extract agar. $y = 12e^{-0.023D} \times 10^7$; D = depth. (Right) Numbers determined by direct count. $y = 3.4 - 0.024D \times 10^9$. Note change in scale for the two graphs.

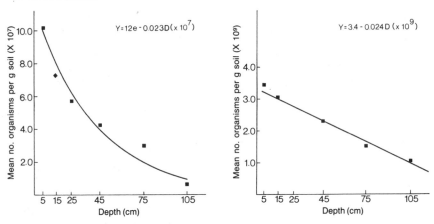

develops a double root system, one near the surface and the second immediately above the water table, the microbial population in the soil layer containing the deep root system is greater than that in soil between the two root-enriched horizons.

Studies of the occurrence of individual species at differing profile depths must rely on cultural procedures or immunofluorescent staining inasmuch as only very rarely are individual bacterial species recognizable in the direct count. *Bacillus* constitutes a greater percentage of the total plate count in the upper than in the deeper profile. Among the fungi, *Chrysosporium* has been found proportionately more numerous at the 30-cm depth than at 10 cm. Most fungal species show preference for the upper profile. Members of the soil fauna that are capable of migration within the profile selectively seek out favorable moisture and temperature regimes.

Distribution of Organisms within the Soil Fabric

Calculations based on conversion of representative biomass carbon values for soil organisms to live biomass and biovolume, and use of median bulk density and porosity values for arable soil, indicate that no more than about 0.2 to 0.4% of the pore space is required for occupancy by organisms. Soil bacteria do not exist as unattached particulates in pores, nor are they easily washed from the soil. Their cells adhere to or are adsorbed on inorganic and organic surfaces. Bacteria show attachment to sand grains by fine fibrillae extending from their cell walls and by extracellular mucilaginous polysaccharides (mucigels). For clustering of sand grains, a combination of microbial and plant mucigels, fungal hyphae, and small rootlets is required (Fig. 5.2).

Most bacteria are larger than clay particles and usually carry a net negative charge but nevertheless become intimately associated with negatively charged clay. Again, fibrillae and mucigels are involved. Mucigels are macromolecules that do not move around in soil, and therefore, it is probable that very fine clays migrate to mucigels and organisms to initiate aggregate formation. Microaggregates are less than 250 μm in diameter, and macroaggregates, more than 250 μm. Pore diameters in microaggregates are 0.2–2.5 μm, and in macroaggregates, largely 25–100 μm. Where open pores exist, pore neck sizes can be assumed to range from equivalence to less than pore diameters. Pore neck size screens entry of organisms, and consequently, invasion of the very small pores in microaggregates is largely by bacteria. As discussed in Chapter 2, either open or closed pores may contain organisms enclosed at the time of aggregate formation. Propagules so enclosed may remain viable over long periods inasmuch as bacteria are known to survive for many decades in stored, dried soils.

Chemical analysis of the soil organic matter (SOM) in microaggregates shows that the contained sugars are mostly of microbial origin. Microbially synthesized sugars are dominantly galactose, mannose, and glucose, whereas those of plant origin contain high proportions of xylose and arabinose. Special staining of ultrathin sections of microaggregates also indicate organic matter of microbial origin. Knowledge that clay–mucigel complexes are strong barriers against diffusion and that the interiors of microaggregates are anoxic suggests that such aggregates are not sites for intensive colonization and metabolic activity by microorganisms.

Macroaggregates permit aeration, water entry and drainage, diffusion of solubles, and occupancy by organisms. Their pores are colonized by microflora and microfauna, and some pore necks may permit entry by members of the mesofauna. Large macroaggregates (>1 mm) may show penetration by roots or close association with them. Macroaggregates are more intensively colonized by organisms and are more active sites metabolically than the soil taken as a whole. Chemical analysis of macroaggregates shows greater content of nutrients (carbon, nitrogen, sulfur, phosphorus) and a greater proportion of sugars of microbial origin than found in the soil generally. Table 5.1 shows the extent to which analytical data for fine clay differ from those for whole soil. It can be concluded that the macroaggregate, with its enhanced nutrient content and hospitable

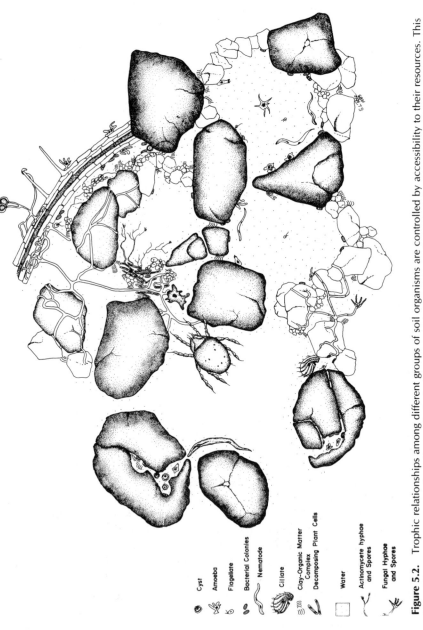

Figure 5.2. Trophic relationships among different groups of soil organisms are controlled by accessibility to their resources. This illustration represents approximately 1 cm² of a highly structured microzone in the surface horizon of a grassland soil. Courtesy of S. Rose and T. Elliott, personal communication.)

Cyst

Amoeba

Flagellate

Bacterial Colonies

Nematode

Ciliate

Clay-Organic Matter
Complex

Decomposing Plant Cells

Water

Actinomycete hyphae
and Spores

Fungal Hyphae
and Spores

Table 5.1
Chemical Analyses of Organomineral Size Fractions from Native Prairie Soils[a]

		mg/g					
Soil	C	N	$P_o{}^b$	Extractable $P_a{}^c$	$\delta^{15}N$ 0/00	C/N	C/P_o
Whole soil	50	4.0	0.58	0.22	11	12.5	87
Sand[e] (>50 μm)	27	1.9	0.19	0.06	7	14.0	141
Coarse silt (50–5 μm)	44	3.6	0.47	0.17	9	11.2	86
Fine silt (5–2 μm)	85	7.9	1.17	0.43	10	10.8	73
Coarse clay (2–0.2 μm)	82	8.7	1.67	0.56	12	9.5	51
Fine clay (<0.2 μm)	63	8.0	1.16	0.37	17	7.9	54

[a]From Tiessen et al. (1984).
[b]P_o, organic phosphorus.
[c]P_a organic phosphorus extractable in sodium bicarbonate and hydroxide.
[d]$\delta^{15}N$ abundance of nitrogen-15 relative to air, as parts per thousand excess.
[e]The sand fraction also includes coarse, floatable (1.00 g cm^{-3}) organic matter.

microclimate, provides a domain more favorable for organisms than found in the soil generally. The intensity of the aggregate-associated bioactivity, however, is less than that associated with two other domains, that of the plant root and its immediately adjacent soil and that provided by plant residues when returned to soil.

Association of Organisms with Plant Roots

Plant root systems occupy the soil horizon most heavily endowed with SOM, and the live, senescent, and dead roots provide substrate materials for microbial growth. The sum total of phenomena occurring at or near the root–soil interface has great impact both on plant welfare and on the soil biota. In discussion of this interface we consider the soil adjacent to and under the influence of roots, the root surface itself, and the root interior. These three regions are variously designated. In 1904, Hiltner used the term rhizosphere principally for the area of bacterial growth around legume roots. In following decades the rhizosphere became generally known as the soil region under the immediate influence of plant roots and in which there is proliferation of microorganisms.

Recognition that the root surface itself is a critical site for interactions between microbes and plants led to its designation as the rhizoplane. The epidermal and cortical tissues of roots have been shown to harbor organisms other than symbionts and pathogens. Colonized root tissue is sometimes referred to as the endorhizosphere, but this term is hardly satisfactory inasmuch as *rhizosphere* has become widely used to denote a soil region. For designation of colonized root tissue, either *histosphere* or *cortosphere* would be preferable. Regardless of terminology, the factor primarily responsible for microbial activity in the three zones (exterior, surface, and interior of the root) is the available carbon contained in or emanating from plant roots.

Early work on the rhizosphere was concerned with the numbers of organisms present. R/S ratios (ratio of organism count in rhizosphere soil to count in root-free soil) were determined for different plant species and for single species in different soils, under differing climatic regimes and at different stages of phenology. Total microbial counts were commonly found to be increased 10- to 50-fold in the rhizosphere. Structurally, the rhizosphere is known to harbor proportionately more G$^-$ bacteria *(Pseudomonas, Achromobacter)* and denitrifiers and fewer G$^+$ and Gram-variable forms *(Bacillus, Arthrobacter)*. Increases in the rhizosphere microflora are accompanied by heightened faunal activity, especially in those groups that are grazers on the microflora or on roots.

Substrate Materials Supplied to Organisms by Roots

Analyses of the organic materials found on, in, or near roots reveal a wide assortment of amino, aliphatic, and aromatic acids, and amides, sugars, and amino sugars. In addition to such solubles and diffusibles is the entire gamut of insolubles occurring in the root (e.g., cellulose, lignin, protein) and lost from it by cell exfoliation and root pruning. Nearly all plant rhizospheres can be expected to contain in varying amounts nearly all of the simple sugars and organic and amino acids. Some of the more complex aromatic acids occur only in certain rhizospheres. Likewise, production of certain microbial attractants and repellents is species limited. Asparagus plants but not tomatoes produce a diffusible glycoside toxic to the stubby-root nematode. Many plants are known to produce nondiffusible compounds that are variously biostatic or biocidal to saprobes and root-invading pathogens. Among such phytoalexins are tomatin, allicin, pisatin, and phaseolin, produced by tomato, onion, pea, and bean plants, respectively.

Pathways for release of plant assimilates from roots include leakage or diffusion of molecules across cell membranes, root secretions and extrusions, and losses of cells and tissue fragments during root growth. Root caps and tips are sites of active exudation; they release mucilaginous material as well as root caps and cells. The main root axis releases mostly soluble and diffusible material and some mucigel. Such loss occurs all along the root, both from cell surfaces and at cell junctions. Root mucigel

Table 5.2
Measurements of Rhizodeposition by Plants

Reference	Observations
Shamoot *et al.* (1968)	For seven clovers and grasses, average nonrespiratory loss of carbon to soil equivalent to 29% of root retention
Barber and Gunn (1974)	For barley grown 3 weeks in solution culture, root loss of amino acids and sugars equivalent to 5 to 9% of dry-matter increment of roots
Barber and Martin (1976)	For cereals in unsterilized soil, loss of assimilates from roots equivalent to 18 to 25% of dry-matter increments
Newman (1978)	1–10% of the root weight lost from roots as solubles, occasionally as much as 25%
Helal and Sauerbeck (1984)	At day 25 in maize, 56% of total assimilate in the tops, 31% in the roots, and 13% lost from roots; the 13% loss divided into 11% released as CO_2 and 2%, added to soil; microbes and roots assumed equally responsible for CO_2 loss

consists of polysaccharides synthesized intracellularly and extruded through the cell membrane. Mucigels are highly hydrated, show a fine fibrillar structure in electron micrographs, and contain carboxyl groups that form bondings to clays. Mucigel is the dominant excretory product of roots; in one study of wheat roots, it accounted for 80% of the total carbon lost from roots other than that ascribable to respiration. Sloughed cells and root prunings possess narrower carbon:nitrogen ratios than does mucigel and closely resemble the composition of the source root tissues themselves.

Carbon loss from the root to the soil during plant growth can be measured by exposure of herbage to $^{14}CO_2$ and subsequent determination of the carbon-14 content of roots, rhizosphere soil, and respiratory CO_2. Selected data obtained by this approach are summarized in Table 5.2. The last entry therein shows 11% of the plant assimilate going to microbial respiration, and 2%, to SOM. Assignment of all such soil carbon to microbial biomass carbon indicates a growth efficiency of 27%.

Spatial Relationships of Organisms and Roots

The number of soil organisms at successive distances from the root surface is inversely correlated with increasing distance. For bacteria, most of the steep fall in numbers near the root occurs within the first 5 μm (Table 5.3). Fungal hyphae occur on the root surface, and strands extend ran-

Table 5.3
Numbers of Microbes at Increasing Distance from the Root Surface

Distance from root surface (mm)	Number of types distinguishable	Estimated frequency $[(10^9 \text{ cells cm}^3)^{-1}]$
0–1	11	120
1–5	12	96
5–10	5	41
10–15	2	34
15–20	2	13

domly therefrom for several millimeters. Bacterial coverage of the root surface is usually reported in the range 5–10%. Fungi other than mycorrhizal forms also show sparse coverage. The ratio between the fungal coverage and the bacterial coverage is variable. Single bacteria are often associated with pits in root cell walls, and clusters of bacteria, at cell junctions. Within the mucigel layer, bacteria usually appear as single cells.

Colonization of epidermal and cortical tissue by nonpathogens is sparse in young roots but becomes widespread as roots age. Invasion of the inner cortex occurs without degeneration of the peripheral tissues. Bacteria show longitudinal distribution in the inner cortex and along grooves between adjacent epidermal cells. As root cells die, lacunae or air spaces are formed and become filled with fungal mycelia.

Association of Organisms with Plant Herbage

Plants normally support an abundant leaf surface biota; this association of organisms with herbage is termed the phyllosphere. Intensity of colonization is influenced by climatic and plant factors. High humidity favors, and insolation disfavors, most heterotrophs; both factors favor phototrophs. Plant stems and barks are often colonized by algae and lichens. Broadleaved plants support more organisms on their leaf surfaces than do grasses. Some populations reported for the phyllosphere are shown in Table 5.4. Gram-negative and yellow-pigmented bacteria dominate the bacterial flora, and yeasts, the fungal flora.

Table 5.5 summarizes some observations on the nature and quantity of phyllosphere exudates. Organic carbon measurable in leaf leachates and canopy throughfall indicates the approximate amount of assimilate potentially available to the biota. Measurement of respiratory CO_2 attributable to organisms on leaves is difficult to achieve in the presence of

Table 5.4
Microbial Populations Reported for the Phyllospere[a]

Plant	Microorganisms (fresh-weight basis)
Clover	$1–30 \times 10^8$ g^{-1} (bacteria)
Grass	$0.1–10 \times 10^6$ g^{-1} (bacteria)
Onion	2×10^6 g^{-1} (yeasts)
Coffee	$10–20 \times 10^6$ cm^{-2} leaf surface (bacteria)

[a]From Clark and Paul (1970).

Table 5.5
Amounts and Kinds of Material in Leaf Exudates and Leachates[a]

Investigator	Plant species studied	Observations reported
Greenhill and Chubnall (1934)	*Lolium perenne*	Glutamine guttated
Dalbro (1956)	*Malus sylvestris*	Precipitation washes down 100 g carbohydrates m^{-2} year^{-1}
Tukey *et al.* (1957)	*Phaseolus vulgaris*	7.5 mg carbohydrate per 24 hr obtained by continued artificial leaching, equivalent to 4.8% of the leaf weight
Carlile *et al.* (1966)	*Quercus petraea*	Precipitation washes down 130 g m^{-2} year^{-1} of dissolved organic matter; 90 g of this total carbohydrate, mainly glucose, fructose, melezitose
Ruinen (1966)	*Aloe* sp., *Sansevieria* sp.	Of fatty acids excreted by leaves, 60% by weight acetic acid, 5–15.5% palmitic, 7–29% oleic; smaller amounts of myristic, stearic, linoleic, linolenic acids
Malcolm and McCracken (1968)	*Quercus falcata, Q. virginiana, Pinus palustris*	Precipitation washes down 2 g m^{-2} year^{-1} organic matter; 1 g or more organic acids; 400 mg, reducing sugars; and 100 mg, polyphenols

[a]From Clark and Paul (1970)

both plant respiration and photosynthesis. Calculations based on microbial biomass suggest that less than 1% of the leaf assimilates are metabolized by the phyllosphere biota.

Nonpathogenic organisms occur within the tissues of fruits, stems, and leaves. The coats of seeds prior to release from fruiting structures are sparsely colonized. During seed dispersal, additional organisms, either casual air contaminants or members of the phyllosphere population, become seed-coat occupants. With the onset of germination and establishment of an active seed metabolism, sugars and other organics are exuded. Seed exudates differ from seedling and root exudates; large quantities of polymeric aromatic acids, such as vanillic acid, are present in seed exudates, and lesser quantities of several oligosaccharides are common in seedling and root exudates. Some seed-associated organisms produce auxins, vitamins, and gibberellin-like substances that benefit emergent seedlings. Other organisms are known to produce substances that delay seed germination, and still others, substances that repress plant pathogens.

Association of Organisms with Plant Litter

In natural ecosystems and in most managed forests and grasslands, the major portion of the net primary production goes to litter. Even in grassland, consumption by large herbivores seldom exceeds 50% of net productivity, and in forests, intake of green herbage by fauna is even lower. Production of aboveground litter in most grassland falls in the general range 100–300 g m^{-2} year^{-1}. Production is comparable in chaparral and shrub communities (150–400 g) but higher in forests (200–800 g). Turnover

Table 5.6
Fungal Genera Observed in Grass Litter and in the Underlying Humus[a]

Fungus	Litter	Humus
Alternaria	71	4
Cladosporium	32	2
Mycelia sterilia	70	12
Penicillium	0	22
Sporomiella	0	20
Trichoderma	3	15
Fusarium	10	27

[a]Numbers represent percentages of plates seeded that yielded the fungi.

Table 5.7
Fungal Species (Excluding Mycorrhizal Fungi) Associated with Decaying Pine Needles[a]

Fungus	Living needles	Brown needles	Dark brown needles	Blackish brown needles	Yellowish needles	Yellowish decayed needles	Humus layer
Lophodermium pinastri	a[b]						
Cenangium sp.	a	a					
Desmazierella acicola		a	a	a			
Endophragmia alternata			a	a			
Unidentified F-1051			a	a			
Kriegeriella mirabilis				a			
Marasmius androsaccus			a	a			
Colybia confluens					a		
Penicillium spp.						a	a
Trichoderma spp.							a

[a]From Soma and Saito (1979).
[b]Symbol ''a'' denotes observance in abundance.

time of fine litter is 2 to 4 years, thus the litter accumulation on site is several times the annual deposit.

The initial invaders of aboveground litter are primarily organisms already present in the phyllosphere. Senescent leaves before detachment suffer attack by both microflora and fauna, and in forests, standing dead wood and branches often lose half their carbon through attack by mesofauna and lignolytic fungi before becoming surface litter. As leaves and small twigs or stems fall to the ground, the number of bacteria thereon increases sharply. The bacterial populations of moist litter exceeds that of the phyllosphere by about two orders of magnitude.

The litter biota varies with depth and with stage of decay. Frequency of observance of fungal genera in grass litter as compared to that in the underlying humus is shown in Table 5.6, and fungal succession in decaying pine needles, in Table 5.7. The soil fauna are more active in forest than in grassland or cropland litter, but even in forests, invertebrates metabolize only a minor portion of the litter carbon (Fig. 5.3). Comparison of respiratory and biomass values does not take into account the role of invertebrates in mixing and fragmenting litter, thereby exposing more surface area for microbial attack. Faunal biomass is concentrated in the surface litter and decreases rapidly in the upper few centimeters of the underlying soil.

Figure 5.3. Comparative biomass and CO_2 flux for the microflora and faunal decomposers in a deciduous forest litter. (From Reichle, 1976.)

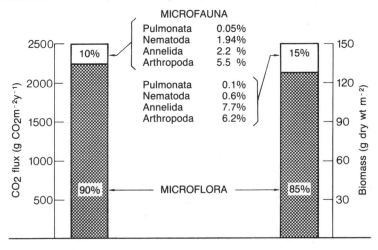

Influences of Management Practices on Soil Organisms

Any management practice that changes soil properties or plant cover affects the soil biota. There are innumerable practices, any of which may variously affect soil water content, temperature, aeration, and pH regimes, and organic carbon and nitrogen levels. Chosen for comment here are the following: soil tillage, addition of pesticides and chemicals, clear-cutting, and controlled burning.

Soil Tillage

Three tillage procedures commonly used are conventional tillage, reduced, minimum, or stubble–mulch tillage, and no-till. The first employs the moldboard plow, usually followed by disking or harrowing. Reduced tillage involves weed control by machinery (sweeps, disks, rod-weeders, chisel plows) that causes only slight incorporation of surface residues and soil disturbance. No-till is weed control by use of herbicides; the only soil disturbance is that associated with seed placement.

Following conventional tillage there is stimulation of microbial activity, as shown by higher census counts and an increase in rate of soil respiration. Such stimulation results from disruption of soil aggregates and better exposure and aeration of their degradable material. Inversion and fragmentation of surface residues fosters a zone of intense microbial activity at plow-sole depth. In contrast, reduced and no-till tillages foster microbial activity at or near the soil surface. In a study in Nebraska, microbial counts at two soil depths were compared for no-till and plowed soil. For the 0- to 7.5-cm depth, counts for total aerobes, facultative anaerobes, and fungi

Table 5.8
Average Ratio of Microbial Populations between No-Till (NT) and Conventional Tillage (CT) at Two Soil Depths[a]

	Ratio NT/CT at	
Microbial group	0–7.5 cm	7.5–15 cm
Aerobic bacteria	1.41	0.68
Facultative anaerobes	1.57	1.23
Actinomycetes	1.14	0.98
NH_4^+ oxidizers	1.25	0.55
NO_2^- oxidizers	1.58	0.75
Fungi	1.57	1.23

[a]From Doran (1980).

were 1.35, 1.57, and 1.57 times higher, respectively, in no-till than in plowed soil. For the 7.5- to 15-cm depth, counts for total aerobes, fungi, aerobic bacteria, and nitrifying bacteria were all significantly reduced in no-till (Table 5.8). Changes in the microfaunal populations following tillage echo changes in the total bacterial count. Tillages and monocultures are generally unfavorable to the mesofauna and macrofauna.

Biocides and Chemicals

Herbicides and foliar insecticides applied at field rates seldom reach the soil in sufficient concentration to cause direct injury to soil organisms. The nitrifying bacteria are the most susceptible, and inhibition of nitrification is occasionally observed following a foliar application of a biocide. Preemergence and foliar herbicides at normal rates have no apparent direct effect on rhizobia, but injury or stunting of the host plant unfavorably affects nodule numbers and nitrogen fixation. Rhizosphere populations of nonpathogenic and asymbiotic organisms may be either increased or decreased, depending on the pesticide employed and its rate of application. Again, the response is determined by changes in the physiology of the plant and particularly by the type and amount of root exudates produced.

Much greater disruption of the soil biota is caused by fungicides and fumigants and other eradicant-type chemicals. Following fungicide or fumigant application to soil at an eradicant rate for the target organisms, there is commonly a depression in the rate of soil respiration for one to several weeks, but after one to several months, there is greater cumulative CO_2 evolution from treated than untreated soil. The greater the microbial kill, the more pronounced the partial sterilization effect on soil respiration.

Almost any fertilizer, if applied in a small area such as a band, may cause local inhibition of microbial activity by imposing osmotic stress. Injections of anhydrous or aqueous ammonia produce biocidal concentrations of ammonia in the center of the injection zone. Nitrification proceeds only at the periphery. Banded urea may also produce concentrations of ammonia that are biocidal to soil organisms.

Clear-Cutting and Controlled Burning

Clear-cutting is widely used in timber harvesting, and it has impact on the soil biota for many years following the year of harvest. Differences in site quality, climate, and plant species involved in revegetation determine both the duration and the severity of that impact. Most studies show marked increase in the biomass of both the microflora and fauna during the first few years, often peaking in the second year and sometimes persisting for as long as 5 to 10 years. Several factors are involved in the doubling or tripling of microbial biomass following clear-cutting. Residual slash is of some importance, but comparisons between slash-cleared and slash-remaining plots show that the slash residue is only partly responsible (Sundman *et al.*, 1978). Availability of dead roots for decomposition, greater susceptibility to decomposition of residual litter because of more favorable moisture and temperature regimes, and accretion of new and easily decomposable litter from regrowth herbs and shrubs are all variously involved. Paralleling the microbial proliferation are increases in soil respiration and nitrogen mineralization. There is zero or near-zero loss of nitrogen in the stream runoff from undisturbed forests; following clear-cuts, there are sizeable losses of nitrogen and other nutrients to stream runoff. Lists of organisms present after clear-cutting mirror those made prior to disturbance, but the numerical dominance of individual species differs greatly (Lundgren, 1982).

Controlled burning is a procedure used for forest rejuvenation and for land clearing in conjunction with shifting agriculture. During and immediately following fire, in contrast to clear-cutting, both the microflora and the fauna are diminished. Both groups recover rapidly as new plant growth and litter accumulation occur. The microflora reseeds itself from underlying mineral soil and from wet and dry deposition, and the fauna, from small islands of litter that escape burning.

References

Barber, D. A., and Gunn, A. B. (1974). The effect of mechanical forces on the exudation of organic substrates by the roots of cereal plants grown under sterile conditions. *New Phytol.* **73**, 39.

Barber, D. A., and Martin, J. K. (1976). The release of organic substances by cereal roots into soil. *New Phytol.* **76**, 69–80.

Clark, F. E., and Paul, E. A. (1970). The microflora of grassland. *Adv. Agron.* **22**, 375–435.

Doran, J. W. (1980). Soil microbial and biochemical changes associated with reduced tillage. *Soil Sci. Soc. Am. J.* **44**, 765–771.

Helal, H. M., and Sauerbeck, D. R. (1984). Influence of plant roots on C and P metabolism in soil. *In* "Biological Processes and Soil Fertility" (J. Tinsley and J. J. Darbyshire, eds.), pp. 175–182. Martinus Nijhoff/Dr. W. Junk, The Hague.

Lundgren, B. (1982). Bacteria in a pine forest soil as affected by clear-cutting. *Soil Biol. Biochem.* **14**, 537–542.

Newman, E. I. (1978). Root microorganisms: Their significance in the ecosystem. *Biol. Rev. Cambridge Philos. Soc.* **53**, 511–554.

Reichle, D. E. (1976). The role of invertebrates in nutrient cycling. *Ecol. Bull.* **25**, 145–154.

Shamoot, S., McDonald, I., and Bartholomew, W. (1968). Rhizodeposition of organic debris in soil. *Soil Sci. Soc. Am. Proc.* **32**, 817.

Soma, K., and Saito, T. (1979). Ecological studies of soil organisms with reference to de-
    composition of pine needles. *I. Rev. Ecol. Biol. Sol* **16,** 337–354.
Sundman, V., Huhta, V., and Niemela, S. (1978). Biological changes in a northern spruce
    forest soil after clear-cutting. *Soil Biol. Biochem.* **10,** 393–397.
Tiessen, H., Stewart, J. W. B., and Hunt, H. W. (1984). Concepts of soil organic matter
    transformations in relation to organomineral particle size fractions. *In* "Biological Pro-
    cesses and Soil Fertility" (J. Tinsley and J. W. Darbyshire, eds.), pp. 287–295. Martinus
    Nijhoff/Dr. W. Junk, The Hague.

Supplemental Reading

Burges, A., and Raw, F. (1967). "Soil Biology." Academic Press, London.
Clark, F. E. (1969). Ecological associations among soil microorganisms. *Soil Biol.* **9,** 125–
    161.
Gray, T., and Parkinson, D., eds. (1967). "The Ecology of the Soil Bacteria." Liverpool
    Univ. Press, Liverpool.

Chapter 6

Carbon Cycling and Soil Organic Matter

Introduction

Microorganisms are nature's garbage-disposal agents. They convert the carbon in organic materials to CO_2 and thereby complete the biological carbon cycle that was initiated during photosynthesis. The quote, "From ashes to ashes, from dust to dust," describes the never-ending search by microorganisms for the energy tied up in the C—H bond. This is the driving force behind nearly all of the nutrient cycling reactions involving organic compounds in soils and sediments. The carbon transfers and transformations at the soil or even the cell level also delineate the other nutrients, such as sulfur, nitrogen, and phosphorus, that become available to plants and microorganisms. Recently, people have realized that global carbon cycles are closely tied to biological productivity and soil organic matter (SOM) turnover. The burning of fossil fuels, extensive forest fires, and intensive soil cultivation have increased atmospheric CO_2 to levels that could have major climatic effects. A discussion of carbon cycling in soils can effectively be introduced from the view of the global carbon cycle. The two are closely related.

Global Carbon Cycle

Estimates of the global carbon reservoirs (Table 6.1) show that before 1860, the air contained about 260 ppm CO_2; more recently, it contained 360 ppm. The carbon in the land biota is slightly less than that in the atmosphere [500×10^9 megagrams (Mg)], whereas the carbon in soils is approximately 1500×10^9 Mg. The continual biological interchange

Table 6.1
Global Carbon Reservoirs and Net Carbon Flux between Major
Reservoirs

Carbon reservoirs	ppm	10^9 Mg carbon[a]
Atmosphere		
1850	260	560
1890	290	630
1986	360	760
Ocean		
Carbonates		20×10^6
Dissolved organics		600
Particulate and sediment organics		3,000
Land		
Biota		500
Humus		1,500
Fossil fuel		10,000
Source and flux size per year (1987)		
Release by fossil-fuel combustion		7
Land clearing		3
Forest harvest and decay		6
Forest regrowth		−4
Net uptake by oceans (diffusion)		−3
Annual flux		9

[a] 1 Mg = 10^6 g = 1 metric ton.

between atmospheric CO_2, land plants, and soil constitutes a very dynamic part of the global carbon cycle.

Total annual plant production in the ocean is estimated as 40×10^9 Mg, and on land, 70×10^9 Mg. Decomposition of the plant residues on a yearly basis nearly equals plant production, but over a time period small differences can lead to major accumulations or deficits. Since the late 1800s, the steady state has been altered. Forest clearing, especially that involving fires, as in the Amazon basin, and the breaking of extensive areas of virgin land in the Americas, Siberia, Australia, and Africa have led to the net transfer of carbon from the terrestrial system to the atmosphere. Table 6.1 shows that this is only partially offset by forest regrowth on abandoned fields and on previously logged areas and by the net flux of CO_2 to the oceans by diffusion.

Burning of coal and oil, which results in input of fossil fuel carbon into the atmosphere at a rate of about 7×10^9 Mg carbon per year, has a major influence on global atmospheric CO_2 levels. Although this is only one-tenth of the annual terrestrial biological cycle, if continued it can be expected to increase global CO_2 to problem levels. When the atmosphere CO_2 reaches 600 ppm, the rise in atmospheric temperature due to the greenhouse effect is predicted to be 6–8°C at the earth's poles and negligible in the tropics. Polar warming could lead to extensive melting of the ice caps and flooding of many coastal cities. A warming trend could also lead to changes in rainfall patterns and thus to new areas of desertification. Beneficially, it could lead to an extension of agriculture to much higher latitudes (further north and south). Also, higher CO_2 concentration should lead to higher photosynthetic and water-use efficiency and to better functioning of plant–microbe associations, such as those occurring in symbiotic nitrogen fixation and in mycorrhizas.

Constituents and Turnover of Organic Residues

The largest fraction of all organic carbon entering the soil is that contributed by plant residues. Plants contain 15–60% cellulose, 10–30% hemicellulose, 5–30% lignin, and 2–15% protein. Soluble substances, such as sugars, amino sugars, organic acids, and amino acids, can constitute 10% of the dry weight. They are readily leached from plant residues and are quickly utilized by soil organisms. It is difficult to separate the decomposition of plant residues from that of SOM. By definition, fine identifiable residues and microbial bodies are part of the SOM fraction and are jointly covered here. Reviews on plant residue decomposition and SOM formation have been given by Stevenson (1982a), Flaig *et al.* (1975), and Tate (1987).

Carbohydrates

The common monomeric sugars in nature (Fig. 6.1) are combined in a number of configurations in plants, soils, and microorganisms. Starch, the major food reserve of plants, contains two glucose polymers, amylose and amylopectin. Amylose consists of an unbranched chain of glucose units joined by $\alpha(1\rightarrow4)$-glucosidic bonds (Fig. 6.2). Amylopectin has the same general structure, with side chains attached by $\alpha(1\rightarrow6)$ bonds. Many bacteria and fungi hydrolyze starch by producing extracellular enzymes collectively known as amylases. α-Amylase reduces both amylose and amylopectin to units consisting of several glucose molecules; β-amylase reduces amylose to maltose. Amylopectin is broken down to a mixture of maltose and dextrins.

Cellulose is the most abundant constituent of plant residues, often being associated with hemicellulose and lignin. It occurs in a semicrystalline state with a molecular weight of 10^6 and is composed of glucose units with

Figure 6.1.   Structural formulas (Haworth projection) of some common sugars.

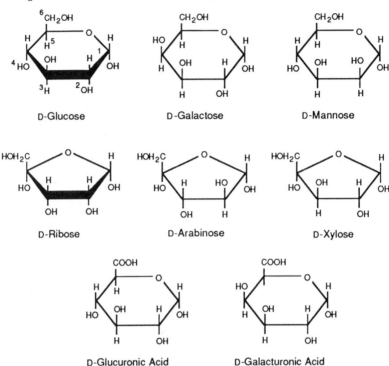

β(1→4) linkages (Fig. 6.3). The individual unbranched chains are held together by hydrogen bonds, as shown in Fig. 6.3b. Decomposition occurs via cellulases from a variety of bacteria, including *Pseudomonas, Chromobacterium, Bacillus, Clostridium, Streptomyces,* and *Cytophaga,* and fungi such as *Trichoderma, Chaetomium,* and *Penicillium.* The term cellulase describes an enzyme complex that acts in two distinct stages. First, there is a loss of the crystalline structure, and then the depolymerization itself occurs. The resultant disaccharide, cellobiose, is hydrolyzed by the enzyme cellobiase to glucose. This can be absorbed by the decomposer or it can enter the soluble carbon pool. Some bacteria, e.g., *Clostridium,* can decompose cellulose anaerobically.

The heterogeneous group of compounds collectively called hemicelluloses are various polymers of hexoses, pentoses, and sometimes, uronic acids. Commonly occurring monomers of the group include xylose and mannose (Fig. 6.1). In the pure state, hemicelluloses are easily decomposed. In nature they are frequently complexed with other substances

Figure 6.2. Structure of starch, showing (a) α(1 → 4) linkage of amylose and (b) α(1 → 6) branched linkage of amylopectin.

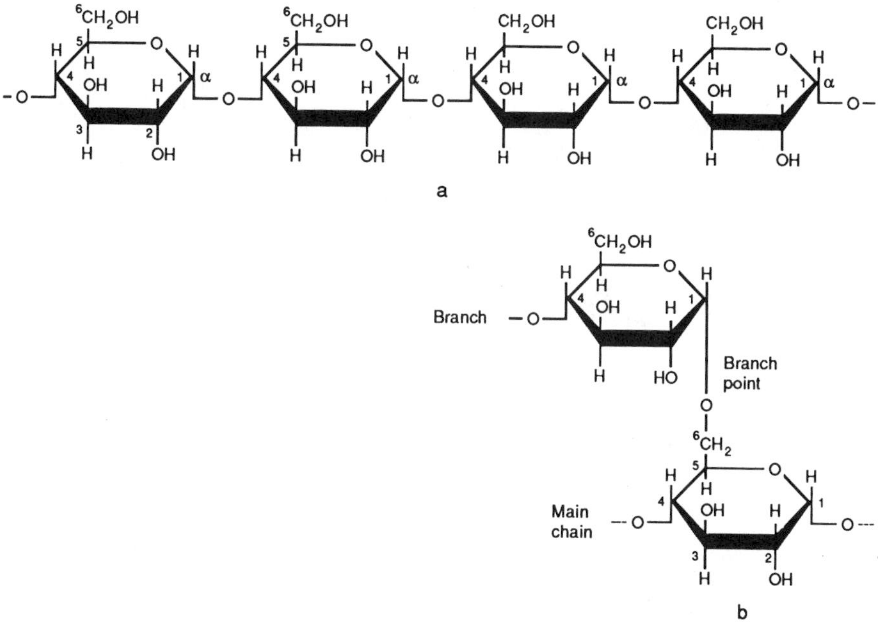

that may make the breakdown more difficult. The microbial attack of the hemicellulose pectin (mostly polygalacturonic acid) has been studied because of the importance of this substance in the middle lamellae of plant cell walls. The process involves several enzymes collectively known as pectinases, which may be repressed by high sugar levels in fresh substrates. In plant tissue degradation, fungi appear to initiate activity but actinomycetes can maintain the attack over a more prolonged period. Pectinases are probably involved in the penetration of host cells by *Rhizobium* and mycorrhizas.

Soil carbohydrates are not an integral portion of the humic acid core. They may be attached as peripheral side chains but more often occur as free polysaccharides. In the free state, they are readily degradable but nevertheless account for 15% of the soil carbon. Reasons for their persistence within the soil must be found. Adsorption of clays and interaction with polycationic metals, such as iron, aluminum, and copper, have been found to greatly increase the resistance of polysaccharides to microbial attack. Another possible reason for persistence is that polysaccharides are the major factor involved in soil aggregation. Polysaccharides present

Figure 6.3.   Cellulose structure, showing (a) β(1 → 4) linkages and (b) hydrogen-bond cross-links between two parallel chains of glucose residues. The chair configuration is the sugar form that probably occurs in nature.

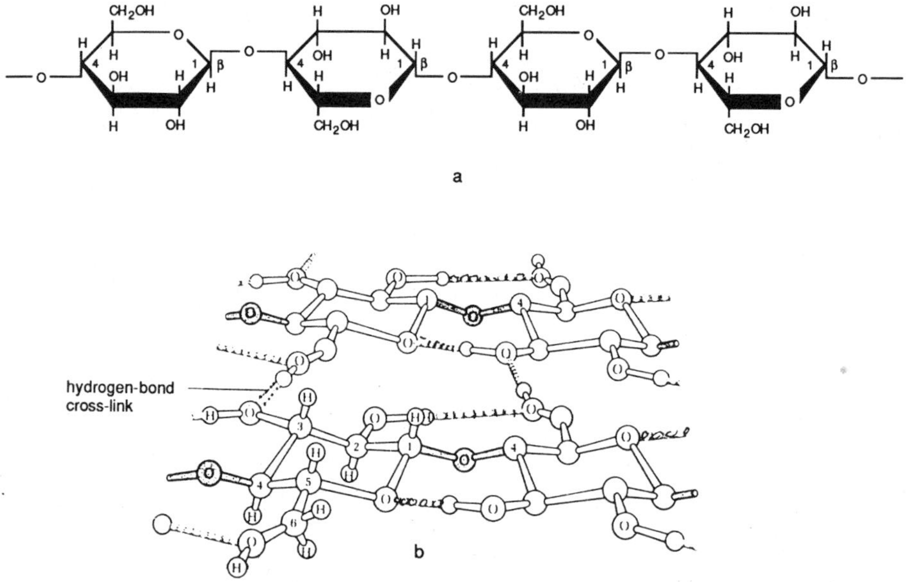

a

hydrogen-bond
cross-link

b

in the middle of aggregates would be prevented from further decay until cultivation or other forces expose the surfaces to microbial attack.

Lignin

The structure of lignin (Fig. 6.4) is based on the phenyl propanoid unit, which consists of an aromatic ring and a three-carbon side chain. Formed by polycondensation, lignin, like aromatic SOM, is not formed by a specific enzyme but in a chemical reaction involving phenols and free radicals. The material therefore does not show a specific order. It is formed as an encrusting material on the cellulose and hemicellulose matrix. Research on SOM indicates that a portion of the lignin entering the soil does not undergo complete decomposition but reacts with microbially produced products in the formation of soil organic matter.

The decomposition of lignin is primarily attributed to fungi. The color of the decayed substrate is indicative of the mode of attack. White-rot fungi are the most active lignin-degrading microorganisms, resulting in the degradation of all wood components to CO_2 and H_2O. Lignolytic white-rot fungi are often known by a number of names. *Coriolus versicolor*

Figure 6.4. Generalized lignin structure, showing the common functional group.

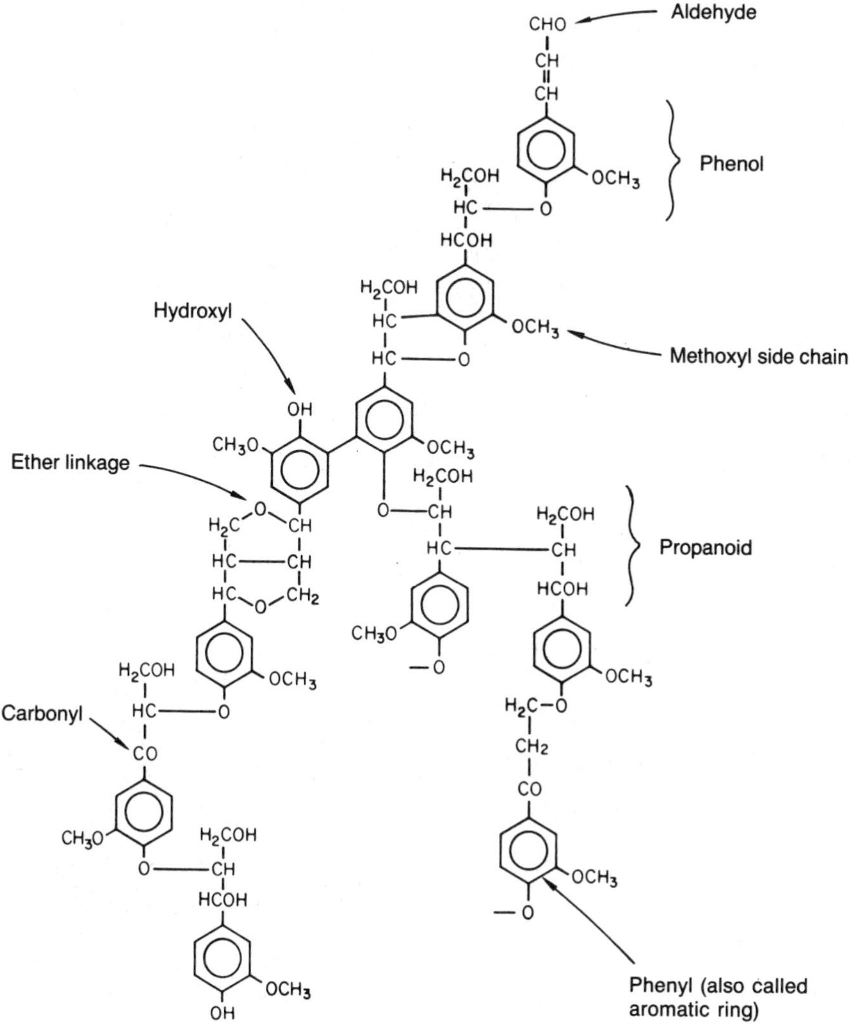

(sometimes included in the genera *Trametes, Polyporus,* or *Polystictus*) decompose ring, methoxyl, or longer side-chain components. *Pleurotus ostreatus* and *Phanerochaete chrysosporium* also cause complete degradation. It is of interest that these fungi are thought to degrade lignin only in the presence of some other readily degradable substrate as the primary energy source. White-rot fungi degrade lignin actively only in the presence of adequate O_2, and it is of interest that high available nitrogen inhibits the lignin degradation. Very little of the lignin carbon is incorporated as cell constituents by lignolytic fungi. It has been shown that on incubation, lignin ^{14}C enters the soil humates but is not found in the soil biomass to any significant extent.

Brown-rot fungi degrade the polysaccharides associated with lignin and remove the CH_3 subgroups and $R{-}O{-}CH_3$ side chains; this leaves phenols behind, which on oxidation turn brown. Representative organisms include *Poria* and *Gloeophyllum*. Another group of fungi, the soft-rot fungi, are important in wet situations and appear to degrade hardwood lignins more effectively than softwoods, with *Chaetomium* and *Preussia* being representative organisms.

Actinomycetes, such as *Streptomyces* and *Nocardia*, and aerobic G^-

Figure 6.5. Decomposition of specifically ^{14}C-labeled benzoic and caffeic acids, caffeic acid linked into phenolic polymers, and several simple organic compounds in Greenfield sandy loam. (From Haider and Martin, 1975).

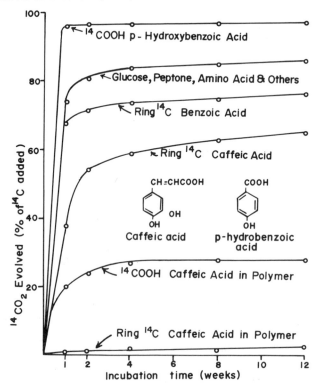

bacteria, such as *Azotobacter* and *Pseudomonas,* lower the lignin molecular size. Whether they can cause complete decomposition and whether they are able to use any of the lignin carbon for growth have not been definitely established. Possibly they attack lignin in order to remove a barrier sheltering cellulose and hemicellulose in the plant residues.

The turnover rate of aromatics in nature is a major factor in determining SOM dynamics. For studying the degradation of aromatics, lignocelluloses containing carbon-14 in either the lignin or cellulose component have been prepared either chemically or by feeding plants with ^{14}C-labeled phenylalanine, cinnamic acid, or glucose. On incubation, rates of CO_2 evolution from the [^{14}C] cellulose component increased rapidly after short lag periods. The degradation rate for the lignin showed a long lag period and was one-fourth that of the associated cellulose, indicating that even materials structurally closely associated decompose at different rates.

The coniferyl alcohols, ferulic acid and caffeic acid, have structures similar to lignin subunits. Studies of their degradation yield useful information concerning residue decomposition. The [14]COOH group of many aromatics is removed by decarboxylation, with little or no incorporation into soil humic constituents (Fig. 6.5). Figure 6.5 also shows the relative degradation and stabilization rates for glucose, simple amino compounds, and the ring carbon-14 of aromatics. Incorporation of caffeic acid into humic-acid-like polymers stabilized all the ring carbon, but the [14]COOH was still preferentially decarboxylated.

Nitrogenous Compounds

The organic nitrogen content of cultivated soil generally parallels that of organic carbon, with the carbon/nitrogen ratio being 10. Of the total soil nitrogen, at least one-third remains structurally unidentified. Acid hydrolysis (boiling in 3 or 6 N HCl for 12 hr) releases half or more of the total nitrogen as amino acids and amino sugars (Table 6.2). Because they are rich in nitrogen, amino acids comprise about 20% of the soil carbon but 30–40% of the soil nitrogen. They account for an even larger percentage of the organic nitrogen that is mineralized annually. Both the total and relative amount of amino acid tend to decrease with cultivation, whereas the relative amount of amino sugar nitrogen stays constant even though the total amount of soil nitrogen declines. This indicates that the amino sugars do not contribute as much as the amino acids to soil nitrogen mineralization. Clay-fixed NH_4^+ described in Chapter 8, is released during acid hydrolysis and constitutes part of the ammonia fraction shown in Table 6.2.

The chemical structures of some of the most common nonsulfur amino

Table 6.2
Nitrogen Distribution in Soils[a]

Climatic zone	Total nitrogen (%)	Form of nitrogen (% of total soil nitrogen)				
		Acid insoluble	NH_3	Amino acid	Amino sugar	Hydrolyzable unknown nitrogen
Arctic (6)[b]	0.02–0.16	13.9 ± 6.6	32.0 ± 8.0	33.1 ± 9.3	4.5 ± 1.7	16.5
Cool temperate (82)	0.02–1.06	13.5 ± 6.4	27.5 ± 12.9	35.9 ± 11.5	5.3 ± 2.1	17.8
Subtropical (6)	0.03–0.30	15.8 ± 4.9	18.0 ± 4.0	41.7 ± 6.8	7.4 ± 2.1	17.1
Tropical (10)	0.02–0.16	11.1 ± 3.8	24.0 ± 4.5	40.7 ± 8.0	6.7 ± 1.2	17.6

[a]From Sowden *et al.* (1977).
[b]Numbers in parentheses show number of soils examined.

acids are shown in Fig. 6.6. The general distribution of amino acids is shown in Table 6.3. Although the acidic amino acids (aspartic and glutamic) appear to be preferentially increased in tropical soils, in general the amino acid composition across the range of world soils is remarkably similar.

Amino acids in the free state are rapidly degraded. Their buildup in the soil can be attributed to their being a component of a relatively stable microbial biomass, their interaction with other organic matter and with soil clays, and their incorporation into aggregates. Basic amino acids with more than one positively charged amino group react more readily with other organic constituents, such as reducing sugars and quinones, than do the acidic and neutral amino acids. They also adsorb more readily to negatively charged clays. The predominance of acidic amino acids in tropical soils, where decomposition rates are highest, is in agreement with findings that these amino acids react strongly with multicharged clays, such as goethite, which are high in iron.

The amino acids are joined by peptide bonds (Fig. 6.7) to form the complex structures common to proteins. Proteins constitute the most abundant nitrogen-containing constituents of organisms, and are readily attacked by many soil organisms. Proteolytic enzymes hydrolyze the peptide links. Enzymes such as trypsin produce proteoses by partial hydrolysis. Microbial enzymes, such as pronase and subtilisin, carry out terminal amino acid chain removal and are more powerful in nature than animal and plant proteoses, such as trypsin and papain. They also have been used to hydrolyze protein constituents of SOM under laboratory conditions.

Fibrous proteins with many cross-links, such as keratin, are resistant to microbial attack, although most actinomycetes, such as *Streptomyces,* and some fungi, such as *Penicillium,* can eventually degrade them. The resistance to decay is due to the presence of disulfide bonds between the

Figure 6.6. Chemical structures of some protein amino acids.

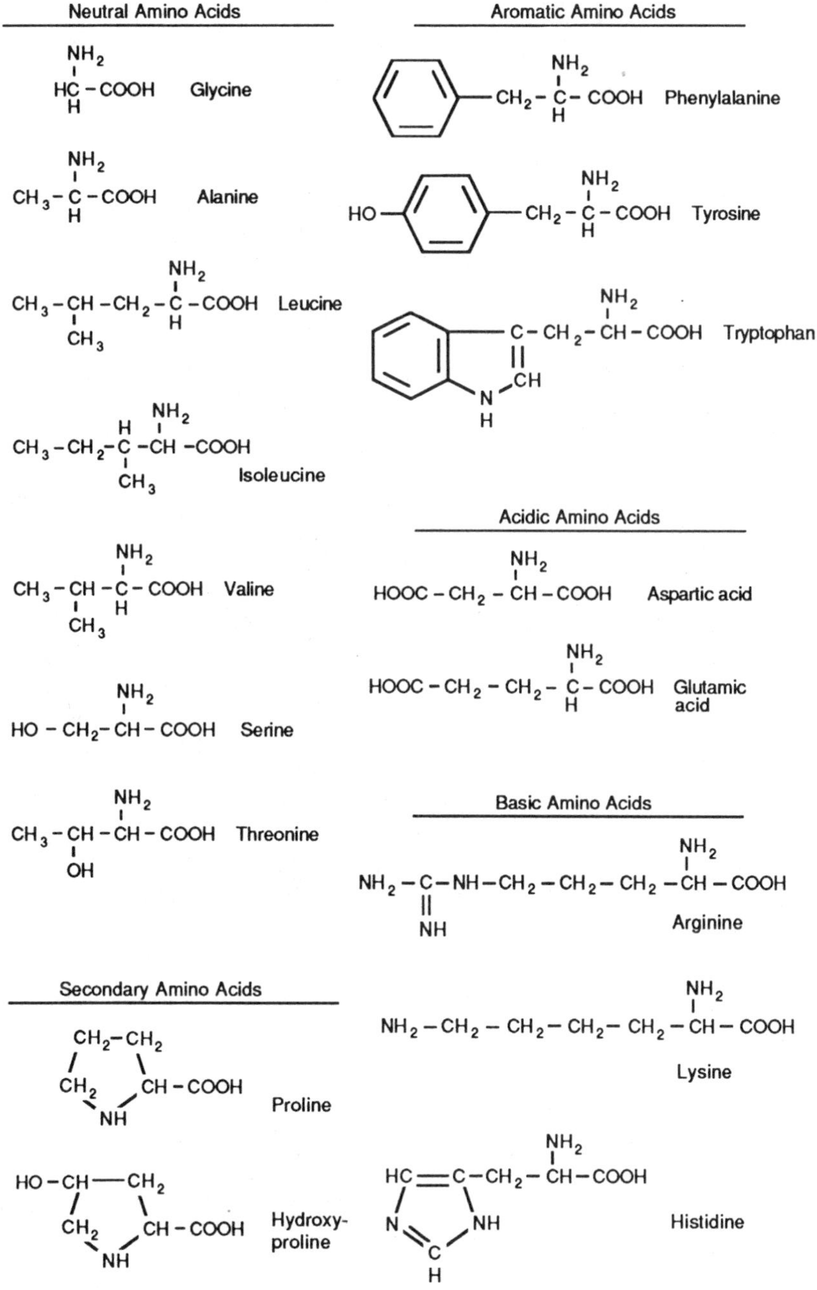

___Neutral Amino Acids___

$$NH_2$$
HC–COOH   Glycine
H

$$NH_2$$
CH_3–C–COOH   Alanine
H

$$NH_2$$
CH_3–CH–CH_2–C–COOH Leucine
CH_3 H

$$NH_2$$
H
CH_3–CH_2–C–CH–COOH
CH_3 Isoleucine

$$NH_2$$
CH_3–CH–C–COOH Valine
CH_3 H

$$NH_2$$
HO–CH_2–CH–COOH Serine

$$NH_2$$
CH_3–CH–CH–COOH Threonine
OH

___Secondary Amino Acids___

CH_2–CH_2
CH_2 CH–COOH
NH   Proline

HO–CH——CH_2
CH_2 CH–COOH   Hydroxy-
NH   proline

___Aromatic Amino Acids___

$$NH_2$$
—CH_2–C–COOH   Phenylalanine
H

$$NH_2$$
HO——CH_2–C–COOH   Tyrosine
H

$$NH_2$$
C–CH_2–CH–COOH   Tryptophan
CH
N
H

___Acidic Amino Acids___

$$NH_2$$
HOOC–CH_2–CH–COOH   Aspartic acid

$$NH_2$$
HOOC–CH_2–CH_2–C–COOH   Glutamic
H   acid

___Basic Amino Acids___

$$NH_2$$
NH_2–C–NH–CH_2–CH_2–CH_2–CH–COOH
NH
Arginine

$$NH_2$$
NH_2–CH_2–CH_2–CH_2–CH_2–CH–COOH
Lysine

$$NH_2$$
HC=C–CH_2–CH–COOH
N   NH
C   Histidine
H

Table 6.3
Distribution of Amino Acids (%) in Major Soil Climatic Zones

Amino acids	Arctic	Temperate	Subtropical	Tropical
Acidic	16	21	22	26
Basic	26	24	25	20
Neutral	54	50	50	51
Sulfur containing	3	3	2	2
Other	1	2	1	2

many cysteine molecules present in keratin. When these bonds are broken by autoclaving, ball milling, or microbial action, the molecule is susceptible to many proteolytic enzymes and is readily degraded.

Although individual amino acids and proteins in plant residues are rapidly degraded, microbially produced organic nitrogen is stabilized in soil. This is due to a relatively slow turnover of the microbial biomass itself and to stabilization of the organic nitrogen by association with resistant organics or within soil aggregates. The nitrogen incorporated in organic constituents often persists in the soil for lengthy periods before being remineralized. The soil nutrients that are turned over at moderate rates, e.g., from 0.5 to 10 years, are said to be in an active fraction of SOM. Those materials that persist for longer periods are defined as being present in old or resistant fractions.

Figure 6.7. Formation of a peptide bond in the beginning of protein formation.

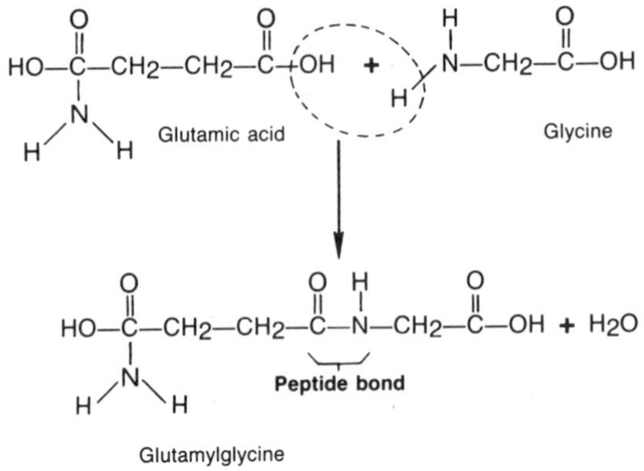

Cell Walls of Organisms

The cell walls of fungi and bacteria are major precursors to SOM formation. Decomposition of fungal cell walls is brought about by genera such as *Streptomyces, Pseudomonas, Bacillus,* and *Clostridium.* Fungi such as *Mortierella* are important under more acidic conditions. Chitin, one of the major contributors of amino sugars in soil, is a major component of fungal cell walls, where it is associated with a number of other fibrous carbohydrate constituents. The acetylglucosamine structure of chitin (Fig. 6.8) contains nitrogen, and its degradation is unlikely to be limited by nitrogen deficiency. Its degradation rate, in addition to being governed by environmental factors, is influenced by the materials with which it is associated. The cell walls of many phycomycetes contain fibers of cellulose, whereas those of organisms such as *Fusarium* contain polymers of the sugars $\beta(1\rightarrow3)$- and $\beta(1\rightarrow6)$-glucan. Unless these hyphae are exposed to both chitinase and glucanase they will not be degraded. Dark-colored pigments, often referred to as melanin, and formed from sugars, and phenolic compounds provide further protection, thus providing a reason for the lower turnover rates of these constituents.

· Cell walls of bacteria have a rigid layer composed of chains of two sugar derivatives, *N*-acetylglucosamine, and *N*-acetylmuramic acid. These chains are joined to each other by a limited number of amino acids linked through peptide bonds (Fig. 6.9). The peptide cross-linkage is most often composed of L-alanine, D-glutamic acid, mesodiamino pimelic acid, and D-alanine.

The thick wall of Gram-positive (G$^+$) bacteria contains peptidoglycans linked to other wall constituents that include a variety of polysaccharide

Figure 6.8. Structure of chitin, showing the *N*-acetylglucosamine group connected by $\beta(1 \rightarrow 4)$ linkages.

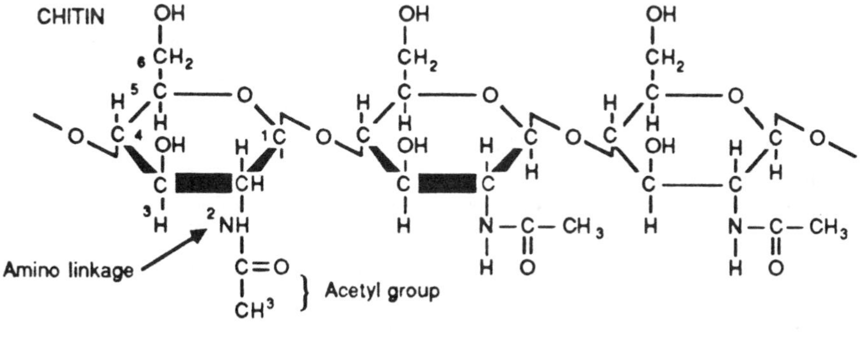

Figure 6.9. Structure of one of the repeating units of peptidoglycan cell wall structure. The structure given is that found in Gram-negative bacteria; other amino acids are found in other bacteria. (From Brock et al., 1984.)

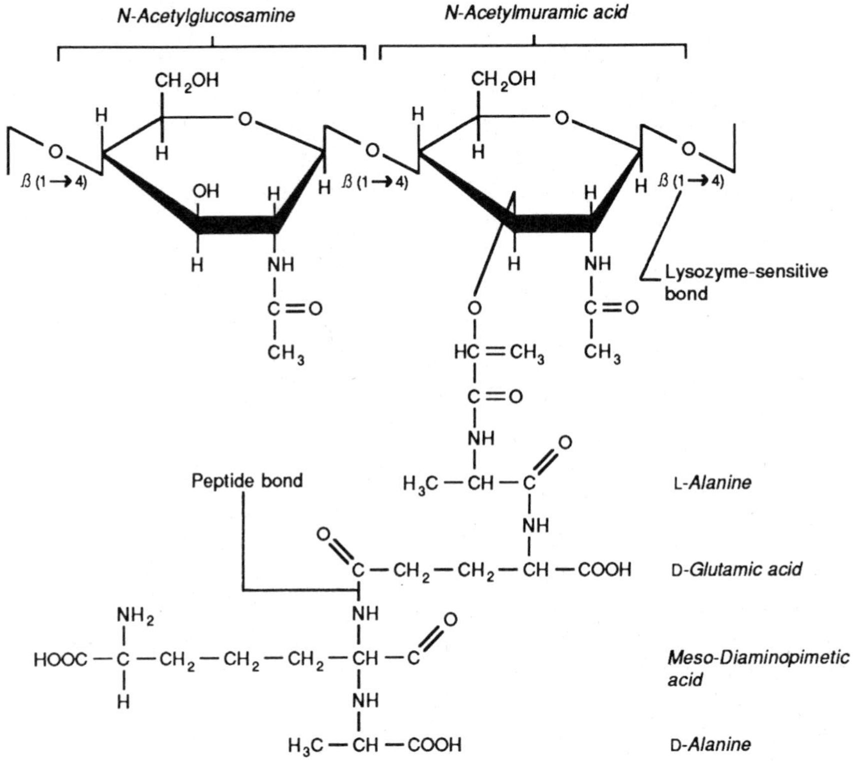

and polyphosphate molecules joined through phosphoid ester linkages. Most contain D-alanine. These materials, which constitute surface antigens, affect the passage of ions through the outer surface layers and are significant in the adsorption of bacteria to clays. Gram-negative bacteria have more complex walls. An outer membrane is composed of lipopolysaccharide, lipoproteins, lipids, etc. A small peptidoglycan layer r is found immediately outside the cytoplasmic membrane. The space between the membrane and the cell wall (the periplasmic space) is an active site for enzyme activity that affects the hydrolytic activity of most soil bacteria.

Microbial growth on a readily available carbon-14 substrate in soil has been found to lead to a build up of residual carbon-14 cell wall materials. These materials are a major but little-studied component of the readily decomposable or soil-active fraction. This fraction supplies more nutrients, such as nitrogen, sulfur, and phosphorus, than more resistant soil fractions. The accumulation of mineralized forms of these nutrients (NH_4^+, NO_3^-, PO_4^{3-}, SO_4^{2-}) during prolonged laboratory incubations is one method of determining the size of this fraction. An alternate technique involves the use of isotope dilution experiments to measure that pool of SOM that is as readily decomposable as microbially produced ^{14}C and ^{15}N compounds formed during extensive incubation of labeled residue in soils.

Roots and Root Exudates

Roots constitute a major carbon reservoir in natural ecosystems and play an important role in continuously cropped systems, where most above-ground material is removed. Some research with crop plants has indicated similar rates of decomposition for above- and beneath-ground materials; other studies have indicated slower rates for roots. The amount of material transported beneath ground varies with plant phenology. Sloughed root materials, such as root hairs and mucilages, constitute major carbon sources, and tracer experiments indicate the total quantity of labeled material in soil is 20–50% higher than that separated as roots at harvest time. This is attributable to sloughed root hairs and other root debris (Lynch, 1983). Root exudates have major qualitative effects and are usually rapidly decomposed, leaving only small amounts of material in extended carbon-14 experiments.

Perennial natural grassland when labeled with $^{14}CO_2$ under field conditions retained 52% of the assimilated carbon above ground and 36% in the shoot bases and roots; only 12% was respired during and immediately subsequent to the labeling period. The carbon-14 data also indicated that more than half the residual root carbon was decomposed in one growing period. Trees have been shown to move large proportions of their photosynthate underground into fine roots. The associated mycorrhizal fungi of some conifers are said to contribute as much carbon to the soil as the aboveground litter.

Formation of Soil Organic Matter

Soil organic matter is composed of decomposing residues, by-products formed by organisms responsible for decomposition of the residues, the microorganisms themselves, and the more-resistant soil humates. Residues of plants differ in phenolic content and in the proportion of lignin to cellulose and protein. Microbial attack of carbohydrates and proteins results in the production of microbial products, and depending on the car-

Figure 6.10. Degradation of plant residues and formation of soil organic matter.

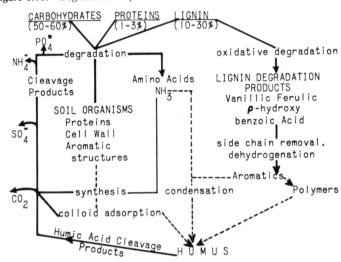

bon:nitrogen:sulfur:phosphorus ratios, in the production of SO_4^{2-} and NH_4^+ as well as CO_2 (Fig. 6.10). There are two physical condensation processes that can explain the production of humates; they are called condensation rather than polymerization because the reactants involved form a series of related but not identical compounds. Since the actual conden-

Figure 6.11. Polyphenol mechanism for the formation of humates, showing the reaction of a phenol (catechol) and an amino compound (glycine).

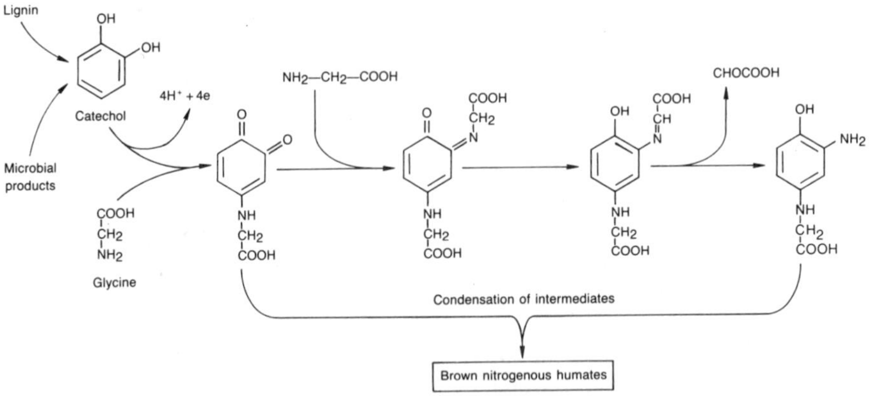

Figure 6.12. Quinone–amino acid complex joining two peptide chains.

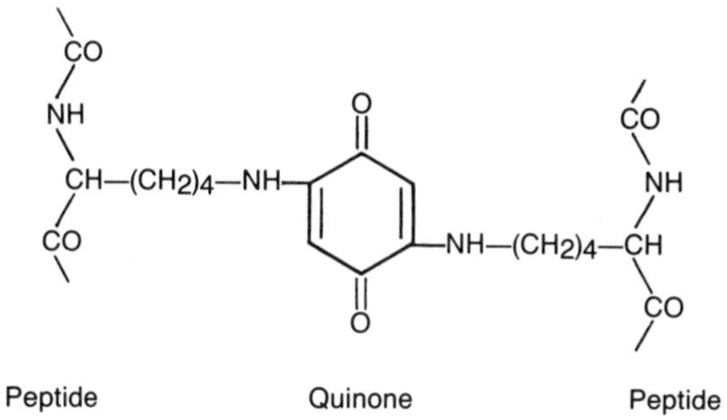

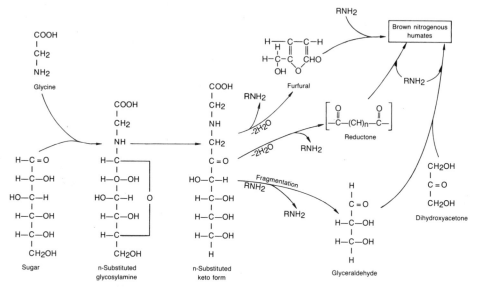

Peptide   Quinone   Peptide

sation of humates is not enzyme controlled, no two of the resultant molecules can be expected to be exactly alike. Yet the general form of humates in most parts of the world is very similar.

One of the two theories for the formation of SOM is shown in Fig. 6.11. This figure shows an amino acid, in this case glycine, reacting with a phenol (catechol) derived either from the partial degradation of lignin or from

Figure 6.13. A generalized mechanism for the formation of humates, showing the browning reaction between the amino acid, glycine, and sugar; other amino compounds or reducing sugars can be involved.

microbial pigments such as those produced by the fungus *Epicoccum*. These form aminoquinone intermediates, which can condense to form brown, nitrogenous humates. The quinone–amino acid or ammonia (NH_3) reaction shown in Figs. 6.11 and 6.12 is considered to be a more significant mechanism of humus formation than the Browning reaction (Fig. 6.13). The browning reaction involves sugars, in this case glucose, reacting with an amino compound such as glycine. The resulting addition compounds rearrange and fragment to form three intermediates: (1) three-carbon aldehydes and ketones, (2) reductones, and (3) furfurals. All these intermediates react readily with amino compounds to form dark-colored end products. The color in both cases is conferred by the series of double bonds involved in the humates. More details are available in the publications by Stevenson (1982a), Flaig *et al.* (1975), and Aiken *et al.* (1985).

Composition of Soil Organic Matter

The carbon of SOM is composed of 10 to 20% carbohydrates, primarily of microbial origin; 20% nitrogen-containing constituents, such as amino sugars and amino acids; 10 to 20% aliphatic fatty acids, alkanes, etc.; with the rest of the carbon being aromatic. Work with nuclear magnetic resonance (NMR) and pyrolysis gas chromatography–mass spectroscopy shows that the content of aromatics in humates of many soils is lower than the 50% previously measured with chemical degradation studies.

Soil organic matter traditionally must be separated from the mineral colloidal matrix of clays and sesquioxides and dispersed in a liquid before being studied. The classical fractionation technique involves dispersion by NaOH or $Na_4P_2O_7$. The fraction not dispersable by the peptizing action of the Na^+, the chelating action of pyrophosphate, and the hydrogen-bond-breaking activity of very alkaline pH values is known as humin. The dispersable material precipitated at acidic pH values is known as humic acids. The material that stays in solution is referred to as fulvic acids (Schnitzer and Khan, 1978).

Fulvic Acids

The materials extracted by NaOH and still soluble at pH 2 are low molecular weight (1000–30,000) fulvic acids. Fulvic acids are composed of a series of highly oxidized aromatic rings with a large number of side chains. Building blocks are benzene carboxylic acids and phenolic acids. These are held together primarily by hydrogen bonding or van der Waals' forces

Figure 6.14.   Schematic diagram of a clay–humate complex in soil. (Adapted from Stevenson 1982a.)

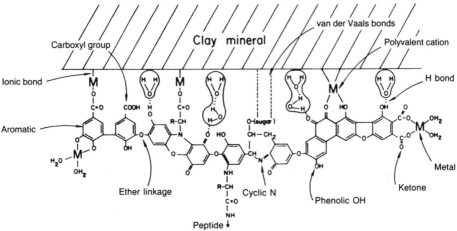

and ionic bonding, as shown for humic acids in Fig. 6.14. X-Ray analysis, electron microscopy, and viscosity measurements of fulvic acids point to a relatively open, flexible structure perforated by voids of varying dimensions that can trap or fix organic and inorganic compounds that fit into the voids, provided the charges are complementary.

Fulvic acids of lower soil horizons that have been produced by leaching, such as those of a spodic B horizon, contain very little nitrogen. Fulvic acids of mollisols, however, contain larger concentrations of nitrogen. This fraction also contains a great deal of polysaccharide materials, as well as low molecular weight fatty acids and cytoplasmic constituents of microorganisms. Their general elemental and functional analyses are given in Table 6.3. These compounds are linear, flexible colloids at low concentrations, and spherical colloids at high solution concentration and low pH values. Their actual shape in nature is determined by their association with each other and with soil inorganic constituents.

Humic Acids and Humin

The organic materials extractable from soil by dilute alkali but precipitable at pH 2 are termed humic acids. These are composed of higher molecular weight (10,000–100,000) materials containing aromatic rings, nitrogen in cyclic forms, and in peptide chains; their general composition is shown in Table 6.4. Since these compounds are formed by polycondensation of

Table 6.4

Elemental and Functional-Group Analysis of Humic and Fulvic Acids

	Elemental Analysis (%)					
Sample	C	H	N	S	O	Ash
Fulvic Acid	49.5	4.5	0.8	0.3	44.9	2.4
Humic Acid	56.4	5.5	4.1	1.1	32.9	0.9

	Functional group analysis (meq g^{-1})			
	OCH$_3$	COOH	Phenolic OH	Total acidity
Fulvic Acid	0.5	9.1	3.3	12.4
Humic Acid	1.0	4.5	2.1	6.6

similar but not identical constituents rather than by enzymes, as is the case for cellulose and protein, no two humic molecules will be identical. The actual structure also has not been determined. There is an increasing amount of evidence from ^{13}C-NMR spectroscopy that much of the SOM is not as aromatic in nature (Aiken *et al.*, 1985) as the structures shown in Fig. 6.14. This shows polyaromatic and nonpolyaromatic building blocks held together by ether linkages, cyclic nitrogen, and hydrogen bonding. Purified humic acids contain few carbohydrate residues. Table 6.4 indicates that the humic acids contain 57% carbon, and there can be 4% nitrogen. The functional groups are primarily COOH groups, phenolic OH groups, alcoholic OH, and a small amount of ketonic oxygen.

The humates are adsorbed to clay minerals by polyvalent cations such as Ca^{2+} and Fe^{3+} (designated as M in Fig. 6.14) and by association with hydrous oxides, either through coordination (ligand exchange) or through anion exchange through positive sites, which exist on iron and aluminum oxides at pH values below 8.

Humin comprises the non-NaOH-dispersable fraction of SOM. Most soils on fractionation give relatively equal amounts of fulvic acids, humic acids, and humin. Removal of the clays in the material not extracted by NaOH, and further extraction, indicates that the humin is composed of fulvic acids plus humic acids in addition to nonsoluble plant and microbial constituents such as undecomposed cellulose, ligniferous materials, microbial cell walls, and some charcoal.

Quantity and Distribution of Organic Matter in Soils

Soil organic matter plays a major role in soil structure and thus has great impact on water penetration, root development, and erosion resistance.

Table 6.5
Global Distribution of Plant Biomass and Soil Organic Carbon

Ecosystem	Area (ha × 10^8)	Net primary production (Mg carbon ha^{-1} year^{-1})	Plant biomass (Mg carbon ha^{-1})	Soil organic carbon (Mg carbon ha^{-1})
Tropical forest	25	19	19	100
Temperate forest[a]	24	12	12	135
Shrubland and savanna	23	8	2	50
Temperate grassland	9	6	0.7	190
Tundra	8	1	0.3	220
Desert scrub	18	1	0.01	60
Rocks and desert	24	0.03	0.5	1
Cultivated	14	6	7	130
Swamp and marsh	2	30		700
	147			

[a]Deciduous plus boreal forest.

It also is the storehouse for the major nutrients, such as nitrogen, sulfur, and phosphorus, and many minor elements. Also, it gives color to the soil and provides cation absorption capacity. It therefore is not surprising that the greatest single component factor controlling the productivity of both cultivated and uncultivated soils is the amount and depth of SOM in the profile.

The quantity of SOM is dependent on the balance between primary productivity and the rate of decomposition. The presence of silt and clay generally increases SOM for a particular set of environmental interactions, and the presence of minerals such as those in certain volcanic soils also results in the greater retention of SOM.

The highest accumulation, 700 Mg carbon ha^{-1} to a depth of 1 m, occurs in highly productive swamps and marshes, where decomposition is inhibited by a lack of O_2 (Table 6.5). Grasslands with wet–dry seasons, and especially if stabilized by Ca^{2+}, have higher organic carbon levels than the equivalent boreal and temperate forests, although their carbon input, as shown by the plant primary production of 6 Mg carbon ha^{-1} year^{-1}, is on the average only half of the 12 Mg carbon ha^{-1} in the forest. Tropical forests, with an annual average net production of 19 Mg carbon ha^{-1},

Table 6.6
Carbon and Nitrogen Contents of Soils under Various Moisture–Temperature Interactions[a]

Temperature–moisture interaction zone	Carbon (Mg ha^{-1} m^{-1})	Nitrogen (Mg ha^{-1} m^{-2})	Carbon/nitrogen
Boreal dry bush	102	6.3	16
Boreal wet forest	150	9.8	15
Boreal rain forest	320	15	22
Cool temperate desert bush	99	7.8	13
Cool temperate grassland	133	10.3	13
Cool temperate wet forest	120	6.3	19
Cool temperate rain forest	200	8.0	25
Warm temperate desert bush	60	3.0	20
Warm temperate moist forest	93	6.5	14
Warm temperate wet forest	270	18	14
Warm temperate rain forest	270	7.0	38
Tropical desert bush	10	0.5	20
Tropical dry forest	100	8.9	11
Tropical wet forest	145	6.6	22
Tropical rain forest	180	6.0	30

[a]Adapted from Zinke et al. (1984).

have less organic carbon than temperate forests, with a productivity of 12 Mg carbon ha^{-1}. This organic matter is distributed deeper throughout the profile, with carbon/nitrogen ratios of 10, similar to that of cultivated soils rather than temperate forests, where the carbon/nitrogen ratio most often ranges from 15 to 20.

Moisture and temperature interactions are known to control the accumulation of SOM. Under moist conditions, the rate of carbon accumulation through photosynthesis is larger than the opposite process of decomposition. A summary of SOM contents from 3600 well-drained soil profiles sampled to 1-m depth and representing the major moisture–temperature interaction zones of the world (Zinke *et al.*, 1984) shows highest carbon accumulation in the wet rain forests of all temperature zones (Table 6.6). Nitrogen contents show an interesting divergence from the carbon data; the accummulation of nitrogen is shown by lower carbon/nitrogen ratios in the dryer areas. The reason for the widening of the carbon:nitrogen ratio of wet areas is not known. It is possible that nitrogen losses are higher in these areas. The possibility of lower nitrogen fixation under the wet conditions also exist.

References

Aiken, G. R., McKnight, D., Wershaw, R., and McCarthy, P. (1985). "Humic Substances in Soil, Sediment and Water." Wiley, New York.

Brock, T. D., Smith, D. W., and Madigan, M. T. (1984). "Biology of Microorganisms." Prentice-Hall, Englewood Cliffs, New Jersey.

Flaig, W., Beutelspacher, H., and Rietz, E. (1975). Chemical composition and physical properties of humic substances. *In* "Soil Components" (J. E. Gieseking, ed.), Vol. 1, pp. 1–211. Springer-Verlag, Berlin and New York.

Haider, K., and Martin, J. P. (1975). Decomposition of specifically carbon-14 labelled benzoic and cinnamic acid derivatives in soil. *Soil Sci. Soc. Am. Proc.* **39**, 657–662.

Lynch, J. M. (1983). "Soil Biotechnology, Microbial Factors in Crop Production." Blackwell, Oxford.

Schnitzer, M., and Khan, S. U., eds. (1978). "Soil Organic Matter." Elsevier, Amsterdam.

Sowden, F. J., Chen, Y., and Schnitzer, M. (1977). The nitrogen distribution in soils formed under widely differing climatic conditions. *Geochim. Cosmochim. Acta* **41**, 1524–1526.

Stevenson, F. J. (1982a). "Humus Chemistry: Genesis, Composition, Reactions." Wiley, New York.

Stevenson, F. J. (1982b). *In* "Nitrogen in Agricultural Soils" (F. J. Stevenson, ed.), *Agronomy*, vol. 22. Am. Soc. Agron., Madison, Wisconsin.

Tate, R. L., III (1987). "Soil Organic Matter. Biological and Ecological Effectors." Wiley (Interscience), New York.

Zinke, P. J., Stanenberger, A. G., Post, W. M., Emanuel, W. R., and Olson, J. S. (1984). "Worldwide Organic Soil Carbon and Nitrogen Data," Publ. 2217. Environ. Sci. Div., Oak Ridge Natl. Lab., Oak Ridge, Tennessee.

Supplemental Reading

Bolin, B., Degens, E. T., Kempe, S., and Ketner, P. eds. (1979). "The Global Carbon Cycle." Wiley, New York.

Chen, Y., and Avnimelech, Y. (1986). "The Role of Organic Matter in Modern Agriculture." Martinus Nijhoff Publ., The Hague.

Ghosh, K., and Schnitzer, M. (1980). Macromolecular structures of humic substances. *Soil Sci.* **129**, 266–276.

Hatcher, P. G., Schnitzer, M., Dennis, L. W., and Maciel, G. E. (1981). Aromaticity of humic substances in soils. *Soil Sci. Soc. Am. J.* **45**, 1089–1093.

Lehninger, A. L. (1982). "Principles of Biochemistry." Worth Publishers, New York.

Sanchez, P., Gichuru, M. P., and Katz, L. B. (1982). Organic matter in major soils of the tropical and temperate regions. *Proc. Int. Soc. Soil Sci. Cong., 12th, 1982,* Vol. 1, pp. 99–114.

Sowden, F. J., Chen, Y., and Schnitzer, M. (1977). The nitrogen distribution in soils formed under widely differing climatic conditions. *Geochim. Cosmochim. Acta* **41**, 1524–1526.

Ziekus, J. G. (1982). Lignin metabolism and the carbon cycle. *Adv. Microb. Ecol.* **5**, 211–243.

Chapter 7

Dynamics of Residue Decomposition and Soil Organic Matter Turnover

Reaction Kinetics Used to Describe Microbial Transformations

Knowledge of the degradation rates of plant residues, microbial bodies, and soil organic matter (SOM) is a prerequisite for understanding the availability and cycling in nature of nutrients such as carbon, nitrogen, sulfur, and phosphorus. Understanding the significance of microbially mediated decay processes involved in SOM decomposition and nutrient cycling can best be accomplished using mathematical analyses of tracer and nontracer data. Enzymes mediate plant residue degradation. Although complex processes are involved, the overall degradation rates can be described by straightforward kinetics. The reaction velocity for degradation can be measured by following the disappearance of substrate components (i.e., plant residue). For kinetic analysis of decomposition, the time interval between each measurement must be short. Many studies of substrate degradation in soil have not used sufficiently short time intervals between measurements and thus have measured values for the degradation of microbial products or of SOM formation rather than for the degradation of the original substrate.

The rate of decomposition (reaction rate) can be expressed as a function of the concentration of one or more of the substrates being degraded. This is termed the order of the reaction and is the value of the exponential used to describe the reaction.

Zero-Order Reactions

Zero-order reactions are ones in which the rate of transformation of a substrate is unaffected by changes in the substrate concentration because

Figure 7.1. Graphic description of zero-order reactions.

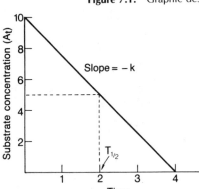 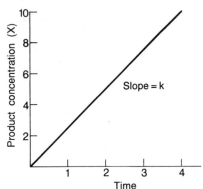

the reaction rate is determined by some factor other than the substrate concentration, such as the amount of catalyst. If a substrate A is transformed to X, the rate of change of A is

$$\frac{dA}{dt} = -k$$

On integration the equation becomes

$$A_t = A_0 - kt$$

where A_t (concentration) is the amount of substrate remaining at any time, A_0 (concentration) the initial amount of substrate in the system, k (concentration time^{-1}) the rate constant, and t (time) the time since the initiation of the reaction.

A useful term to describe the reaction kinetics is the half-life, which is the time required to transform 50% of the initial substrate, e.g.,

$$A_t = A_0/2, \quad \text{then} \quad t_{1/2} = A_0/2k$$

The mean residence time (turnover time) is the time required to transform a quantity of material equal to the starting amount A_0, i.e., $A_t = A_0$; then

$$t_{mrt} = A_0/k$$

At high substrate concentrations, at which substrate levels are not limiting, enzymatic reactions are usually zero order, e.g., nitrification at high NH_4^+ levels and denitrification at high NO_3^- levels. Figure 7.1 shows plots of zero-order reactions.

First-Order Reactions

In first-order reactions, the rate of transformation of a substrate is proportional to the substrate concentration. Consider the reaction

A → X

then the rate of change of reactant A with time is

$$\frac{da}{dt} = -kA$$

The decrease in the concentration of the reactant with time t is dependent on the rate constant k times the concurrent concentration A of the reactant. After integration the following equation is obtained:

$$A_t = A_0 e^{-kt}$$

where A_t is the concentration of substrate remaining at any time t. The rate constant k, which is expressed as inverse time, involves some expression per unit time (e.g., hr^{-1}, day^{-1}, or $year^{-1}$). This equation can be rearranged to facilitate graphic determination of the rate constant. By dividing through by A_0 and taking the natural logarithm of both sides we obtain

$$\ln(A_t/A_0) = -kt$$

A plot of $\ln(A_t/A_0)$ versus t is linear with a slope of $-k$. The equation can also be plotted in log 10, which changes the slope to $-k/2.303$. Figure 7.2 shows typical graphs of a first-order function. The rate constant k (inverse time) is independent of the substrate or product concentration since the slope is constant over time. To calculate the time required to transform one-half the initial substrate ($A_t = A/2$), we start with

$$\ln\left[\frac{(A_0/2)}{A_0}\right] = -Kt_{1/2}$$

which is

$$t_{1/2} = 0.693/k$$

The mean residence time (turnover time) for first-order reactions is equal to $-1/k$. Note that t_{mrt} for zero-order reactions is A_0/k.

Figure 7.2 indicates that plotting $\ln(A_t/A_0)$ versus time t will equal a straight curve if the reaction is first order. Decomposition curves for complex substrates, such as straw, usually yield a multislope decomposition curve. This indicates that the straw consists of several components having

Figure 7.2. First-order function $A = 10e^{-0.5t}$, plotted on linear (left) $\log_{10}$ (center), and $\log_e$ (ln) (right) scales.

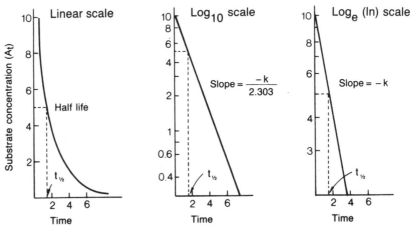

different rate constants. These can be graphically analyzed by curve-splitting techniques, which involve the subtraction of data for the decomposition of the most stable component, such as lignin, from the overall decomposition curve. Plotting these new points will usually yield a straight curve representing k for a component such as cellulose. This can be repeated to obtain the decomposition rate constant for the rapidly degrading proteins, for example. These techniques can be applied only after the values for new product formation, such as microbial growth, are subtracted from the original decay curve. An example of such a calculation is shown in Fig. 7.6. It is now easier to solve a multicomponent curve by computer analysis than by the above graphical method.

Hyperbolic Reactions

Many reactions found in nature have rate constants in which the plot of product (X) versus time yields a curve approaching some maximum value. This is best described by a hyperbolic equation. In enzyme chemistry, the hyperbolic equation is known as the Michaelis–Menten equation. When used to describe microbial growth, the same general function is called the Monod equation. In physical chemistry the use of a curve approaching an asymptote to describe absorption phenomena is often called the Langmuir equation. All three utilize identical principles. The following example is based on the usual discussion for enzyme kinetics employing the hyperbolic function:

$$V = V_{max} \frac{A}{K_m + A}$$

where V is the rate of reaction (concentration time^{-1}), A the substrate concentration (concentration), V_{max} the maximum reaction rate (concentration time^{-1}), and K_m the Michaelis constant (concentration) or "half-saturation constant." The graphic expression for the above equation is shown in Fig. 7.3. Therein V_{max} is the maximum transformation rate; K_m is the affinity or half-saturation constant, which is the concentration A when

$$V = \tfrac{1}{2} V_{max}$$

If the substrate concentration A is much greater than the half-saturation constant K_m, then K_m can be neglected in relation to A and the equation becomes

$$V = V_{max}$$

which describes a zero-order reaction.

At low substrate concentrations, A is less than k_m, and

$$V = kA$$

which describes a first-order reaction, where $k = V_{max}/k_m = $ const.

Figure 7.3. Graphical expression of the Michaelis constant K_m of an enzyme reaction described by the hyperbolic function.

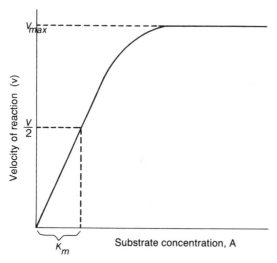

Figure 7.4. Determination of Michaelis constant K_m by a Lineweaver–Burk reciprocal plot.

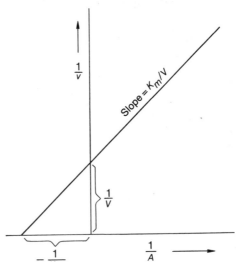

The first-order curve is best solved when plotted as a straight curve. Similarly, hyperbolic curves are best analyzed when they are transformed to linear ones. The well-known double reciprocal plot (Lineweaver–Burk) transforms the hyperbolic equation such that it is linear on ordinary graph paper, as shown in Fig. 7.4. Computers have made the analysis of hyperbolic curves simpler and more meaningful than the transformation to linear graphical equations. It is thus possible to use a number of methods of graphical or mathematical analysis to obtain the most meaningful values for V_{max} or K_m. The great applicability of the use of hyperbolic equations stems from the fact that the whole asymptotic curve can be described by two values, V_{max} and K_m.

Dynamics of Decomposition

Modeling of the dynamics of plant residue or SOM decomposition in the field requires meaningful mathematical expressions for the biological, chemical, and physical processes involved. Decomposition of plant residues has been experimentally found to be reasonably well described by first-order rate kinetics. This means that the decomposition of plant residues is linearly proportional to the plant residue content, but the specific rate constant k is independent of the residue content. It is important to differentiate between the commonly used term decomposition rate, which describes the effect of the overall reaction, and the rate constant k, which is a characteristic of the type of plant residue undergoing decomposition. The use of first-order kinetics to describe the decomposition of SOM implies that the microbial inoculum potential of soil is not limiting the decomposition rate. This is true, in large part, because the high soil microbial biomass often has a fast growth rate relative to the length of most decomposition studies.

Only a portion of the actual decomposition is accounted for when determining the decomposition rate by measuring CO_2 output or the amount of carbon left in the soil. Microorganisms use carbon compounds for biosynthesis, forming new cellular or extracellular material, and as an energy supply. In the latter process, carbon compounds are converted largely into CO_2, microbial cells, and waste products. Under aerobic conditions, the level of waste products usually is not high; therefore, the amount of biosynthesis, or the production of microbial cells, can be calculated from CO_2 data. This requires a knowledge of yield or efficiency of substrate conversion to microbial biomass:

$$C = C_i[1 + Y/(100 - Y)]$$

where C is the substrate decomposed, C_i the CO_2 carbon evolved, and Y the efficiency of the use of carbon for biosynthesis, expressed as a percentage of the total carbon utilized for production of microbial material.

It is difficult to measure the proportion of undecomposed substrate, such as plant residues, remaining in soil. Even when the original compound can be determined chemically, measuring the true decomposition may be

Table 7.1

First-Order Decay Constants (with and without Correction for Microbial Biosynthesis) during Decomposition of Organic Compounds Added to Soil under Laboratory Conditions

Material	Time (days)	k (days^{-1}) Uncorrected	Corrected efficiency % 20	Corrected efficiency % 60
Straw–rye	14	0.02	0.03	0.11
Hemicellulose	14	0.03	0.04	0.11
Lignin	365	0.003	0.006	—
Glucose	1	—	—	4.0
Native grass	30	0.006	0.008	.02
Fungal cytoplasm	10	0.04	0.05	.17
Fungal cell wall	10	0.02	0.03	.07

hampered by microbial production of that particular compound, as in the case of sugars. The decomposition rate constants k, corrected for biosynthesis, differ significantly from the uncorrected ones (Table 7.1). Growth efficiencies of 40 to 60% are generally considered to be realistic for the decomposition of carbohydrates; other compounds, such as waxes and aromatics, result in much lower efficiencies. Where data over extended periods only are available, it is not possible to calculate true decomposition values and microbial growth efficiency because CO_2 is evolved both from the original substrate and from microbial cells produced in the first flush of activity. The data show that one makes a serious mistake when assessing decomposition rates of amendments without accounting for microbial biosynthesis of a portion of the attacked substrate.

Modeling of Plant and Soil Data

Experiments both in the field and the laboratory have shown that the decomposition rate constant k describing the decomposition of plant material is nearly always independent of the quantity added if the carbon addition does not exceed 1.5% of the soil dry weight. Higher addition rates can alter the soil characteristics and slow down decomposition.

Decomposition has been found to depend on plant carbon, nitrogen, lignin, and carbohydrate composition. One method of expressing this dependence involves that used by Herman et al. (1977), where

$$CO_2 \ \text{ evolved} = \frac{\sqrt{\% \text{ carbohydrate}}}{\text{C/N residue} \ \times \ \% \text{ lignin}}$$

After extended incubation periods, the proportion of different plant material retained in various soils under different climatic conditions is very similar and often approaches 20% of the original amount. This is due to a similarity in microbial production and stabilization of soil organic matter even though the initial decomposition rates vary widely. Work with forest litter also has shown that approximately 20% of the carbon in the original litter is found after the initial period of decomposition is over. This includes some resistant lignin products but is composed primarily of microbial products and soil humic constituents.

A useful simulation model describing SOM turnover over long periods of time was published by Jenkinson and Rayner (1977). The data they used came from the long-term plots in Rothamsted, England. These had been sown to wheat for 130 years. The data came from (1) incubation experiments using ^{14}C-labeled plant materials, which provided data over a 1- to 10-year period, (2) radiocarbon dating of SOM, which supplied the apparent age of the old fraction (ranging from 100 to 3000 years), (3) the effect of radiocarbon produced through thermonuclear explosions and distributed throughout the atmosphere during the period 1950–1970, which provided information on material that had been in the soil for 5 to 20 years, (4) chloroform ($CHCl_3$) fumigation incubation method (CFIM) data on microbial biomass, as described in Chapter 3, (5) knowledge of plant input levels, and (6) measured long-term SOM levels in these plots. Five soil carbon fractions were considered: (1) decomposable plant material, $k = 4$ year^{-1}, (2) resistant plant material, $k = 0.03$ year^{-1}, (3) microbial biomass, $k = 0.41$ year^{-1}, (4) physically stabilized organic matter, $k = 0.014$ year^{-1}, and (5) chemically stabilized SOM, $k = 0.0003$ year^{-1}. As shown earlier, a k of 4 year^{-1} represents a half-life of 0.17 year^{-1}, whereas the stabilized SOM has a half-life of 2000 years. The decomposition rate of each of these components was assumed to be first order. It was also assumed that during decomposition each of the five components decayed to CO_2, microbial biomass, physically stabilized SOM, and chemically stabilized SOM in similar proportions.

The fit between the model and the experimentally determined results on these long-term field plots suggested that the model was a useful representation of the turnover of the SOM in cropped soils. The output from the Jenkinson–Rayner model showed a microbial turnover time of 2.5 years, with 34 kg nitrogen ha^{-1} year^{-1} flowing through the microbial biomass annually in these unfertilized, low-yielding wheat plots (Table 7.2).

Use of the Jenkinson–Rayner model for a wheat field in Canada and a sugar-cane soil in Brazil provide very interesting comparisons (Table 7.2).

Table 7.2
Carbon and Nitrogen Turnover in Three Different Ecosystems[a,b]

Determination[c]	Rothamsted	Canada	Brazil
Soil weight (Mg ha^{-1})	2200	2700	2400
Organic carbon (Mg ha^{-1})	26	65	26
Carbon inputs (Mg ha^{-1} year^{-1})	1.2	1.6	13
Turnover of soil carbon (years)	22	40	2.0
Microbial carbon (kg ha^{-1})	570	1600	460
Microbial nitrogen (kg ha^{-1})	95	300	84
Microbial turnover (years)	2.5	6.8	0.24
Nitrogen flux through microbial biomass (kg ha^{-1} year^{-1})	34	53	350
Crop removal of nitrogen (kg ha^{-1} year^{-1})	24	40	220

[a]Adapted from Paul and Voroney (1984).
[b]Western Canadian wheat–fallow, Rothamsted U.K. continuous wheat without fertilizer, and Brazilian sugar-cane soil–plant systems.
[c]Mg, megagram = metric tonne.

The mollisols in Canada, under a wheat–fallow crop rotation, had high reserves of organic carbon and nitrogen but with plant residue inputs similar to those at Rothamsted. The microbial nitrogen determined by using CFIM represented 300 kg nitrogen ha^{-1}. The turnover time of microbial biomass of 6.8 years shows the potential for stabilization of a largely inactive population in soils with high SOM levels.

The spodosol in which the sugar cane was grown in northeast Brazil had carbon and nitrogen contents similar to those of the continuous wheat plots in Rothamsted, but microbial biomass carbon and nitrogen were lower. The high input of carbon from the sugar cane (13 Mg ha^{-1} year^{-1}) resulted in a calculated turnover time of 0.24 for the microbial biomass. This was 28 times faster than that found in the Canadian mollisol and demonstrates the situation in which the biomass acts more as a catalyst and a short-term reservoir than as a major source–sink for nutrients. Although the Brazilian biomass nitrogen represented only one-third of the nitrogen removed by the crop, the estimated nitrogen flux of 350 kg nitrogen ha^{-1} year^{-1} through this biomass was 1.5 times that removed by the crop.

The above models were developed to describe long-term effects of plant production on SOM synthesis. However, a more mechanistic model is required to simulate the degradation of complex materials.

Model Outputs

A model, such as that in Fig. 7.5, can help obtain a better understanding of the quantitative aspects of decomposition of a complex substrate such as crop residues. This understanding is best obtained when one also considers the production of microbial materials from the original substrate. The complex substrate (straw) was considered to consist of three fractions: (1) easily decomposable sugars and amino acids, (2) slowly decomposable cellulose and hemicellulose, and (3) resistant lignin. Knowledge of the decomposition rate constants k and microbial growth efficiencies are required to determine each of the flows, and these are given in Table 7.3. The decay rate constant of the microbial biomass in this model is a function of all processes that require substrate other than for growth. These include maintenance of cell integrity and nutrient uptake, for example. The concept of a separate constant for maintenance energy as developed in chemostats is not readily applied to soil populations, where different subgroups with varying levels of activity occur simultaneously. Measurements of maintenance energy of a soil population that is essentially in a steady state are much lower than values derived from chemostats. Natural populations also feed on each other (cryptic growth). The concepts of maintenance

Figure 7.5. Model describing decomposition of carbon in plant residue decomposition of plant residues and the turnover of soil organic matter constituents. ——, Carbon flows. Numbers in circles refer to pools shown in Table 7.3. (From Van Veen et al., 1984.)

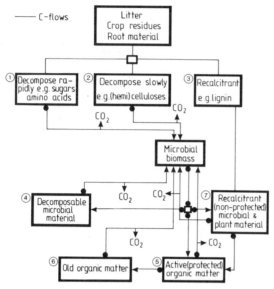

rates and cryptic growth are therefore most easily considered by incorporating them into a death rate constant. Once dead, the microbial biomass can be considered as an easily decomposable pool (cytoplasm) and as more-resistant components (cell walls).

The model output, based on first-order kinetics (Fig. 7.6), shows the

Table 7.3

Pool Sizes, Decomposition Rates, and Efficiency of Microbial Production Used in a Carbon Turnover Model Describing the Decomposition of 1000 μg Carbon g^{-1} of Soil

Pool	Residue carbon ($\mu g\ g^{-1}$ soil)	Decomposition rate k (days^{-1})	Utilization efficiency (%)
Easily decomposable (1)[a]	150	0.2	60
Slowly decomposable (2)	650	0.08	40
Lignin (3)	200	0.01	10
Decomposable microbial products (4)[b]	6	0.8	40
Active protected SOM (5)	5000	3×10^{-4}	20
Old organic matter (6)	7000	8×10^{-7}	20
Recalcitrant plant and microbial products (7)[b]	4	0.3	25

[a]Numbers in parentheses refer to pools shown in Fig. 7.5.
[b]Pool sizes of biomass pools 4 and 7 are much larger. These are initial rates. The pool sizes shown refer to those produced from the added substrate.

Figure 7.6. Decomposition of straw carbon (C) in the laboratory, plotted as a series of first-order reactions after correction for microbial production. The actual plant carbon-C remaining as proteins and solubles (C_1), cellulose and hemicellulose (C_2), and lignin (C_3) is much lower than the total carbon remaining. The equation for actual decomposition shows the initial content C of each component and the decomposition rate k:$A = C_1 e^{-k_1 t} + C_2 e^{-k_2 t} + C_3 e^{-k_3 t}$; $100 = 15e^{-0.2t} + 65e^{-0.08t} + 20e^{-0.01t}$.

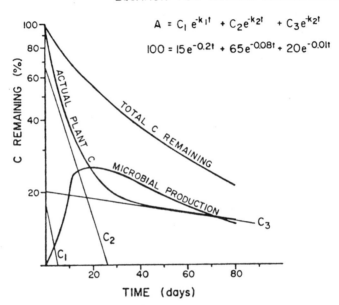

EQUATION FOR ACTUAL DECOMPOSITION

$$A = C_1 e^{-k_1 t} + C_2 e^{-k_2 t} + C_3 e^{-k_2 t}$$

$$100 = 15e^{-0.2t} + 65e^{-0.08t} + 20e^{-0.01t}$$

large difference between the true decomposition after correcting for microbial growth, and the decomposition that would have been measured when only plant carbon remaining in the soil was utilized. Growth efficiencies of 60% are often found for growth of a soil population on simple substrates such as glucose. More complex substrates have lower efficiencies, as shown in Table 7.3. Aromatics such as lignin appear to be largely cometabolized by fungi. This involves the degradation of the substrate by enzymatic systems but little uptake of the breakdown products for the production of energy. The fungi gain little, if any, energy for growth and incorporate little carbon during the decomposition of the aromatics. Therefore, aromatic decomposition occurs only in the presence of other available substrate, such as cellulose.

The model described in Fig. 7.5 and in Table 7.3 is based on decomposition rate constants determined under laboratory conditions and provides output curves such as in Fig. 7.6. Corrections for field moisture and temperature conditions involves the use of correction factors, such as

Table 7.4
Effect of Environment on Decomposition Rate of Plant Residues Added to the Soil

Residue	$T_{1/2}$ (day)	k (days^{-1})	Relative rate
Wheat straw, laboratory	9	0.08	1
Rye straw, Nigeria	17	0.04	0.5
Rye straw, England	75	0.01	0.125
Wheat straw, Saskatoon	160	0.003	0.05

those shown in Figs. 2.6 and 2.9. It has been experimentally determined that this model adequately describes decomposition of residues in the laboratory and in Nigeria, England, and western Canada (Table 7.4). The decomposition rates of 0.08, 0.04, 0.01, and 0.003k (days^{-1}), respectively, for these sites integrate the effects of moisture and temperature. This shows that the average decomposition rate in western Canada, with cold winters and dry summers, is one-twentieth that of the laboratory and one-tenth that of the tropical soil, which has some dry periods.

The residue decomposition rate in these experiments was insensitive to soil type. This can be explained by the fact the straw type was similar and that the larger straw particles formed their own microhabitat. The stabilization of SOM and also the microbial population, however, is soil-

Figure 7.7. Prediction of the effect of rainfall-induced soil erosion, calculated by the universal soil loss equation, on the organic carbon remaining in the surface 15 cm of a cultivated soil in continuous cropping and crop fallow. (From Voroney *et al.*, 1981.)

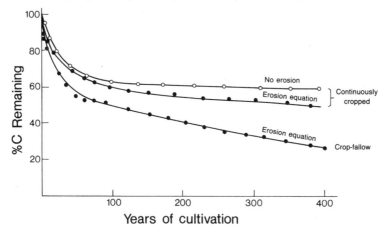

type dependent. Clay type, such as allophane, and soil silt and clay content control stabilization rates.

Mathematical modeling can also prove valuable in predicting long-term changes in SOM from cultivation, and effects on the forest floor after clear-cutting or fire. Figure 7.7 shows the output of a model based on information from carbon dating of the resistant fraction and the determination of plant residue decomposition rates during a 10-year field experiment in western Canada. These data were then related to residue input levels in a long-term SOM decomposition model. The model incorporates the concept of first-order plant residue decomposition rates, microbial growth, and the stabilization of microbial products by soil clays and silts. It predicts the long-term effects of different cultivation practices and erosion. Data such as these can be extremely important in management decisions concerning one of the world's most important natural resources, the soil. They are also applicable in estimating the interaction of SOM and the global carbon cycle on CO_2 contents in the atmosphere and possible climatic changes, as discussed in the introduction of Chapter 6.

Sample Calculations Involving Reaction Kinetics and Soil Microbial Activity

1. There is a *high* substrate concentration of glucose, 2.0% w/w basis, at the start of decomposition. After 2 hr in a mixed culture vessel, the concentration of substrate is 1.6%. The reaction proceeds as follows:

Time (hr)	Substrate remaining (%)
0	2.0
2	1.6
4	1.2
6	?

   (a) What is the rate constant k?
   (b) How much substrate will be left after 2 hr?
       How much is left after 6 hr?
   (c) What is the half-life of the initial glucose level?
   (d) What is the mean residence time?

2. The following data are given for a reaction whose quantity of transformed substrate is proportional to the substrate concentrations:

   At t_0, the initial concentration of X = 100 mmol. The rate constant k is equal to 0.3 hr^{-1}. After 10 hr, what is the concentration of substrate X?

3. Given an initial concentration A_0 of substrate A of 50 mmol in a reaction vessel, and that after 8.5 hr there remains 32 mmol:
   (a) What is the rate constant of this first-order reaction?
   (b) What is its mean residence time?

4. Given a reaction vessel that initially has 60 mmol of substrate, and that after 12 hr it has a concentration of 40 mmol, and after 24 hr, 26.7 mmol:
   (a) What type of reaction is it, first or zero order?
   (b) What is its k (reaction rate)?
   (c) What is the substrate concentration after 6 additional hr?

5. A microbiologically mediated reaction has the following characteristics:

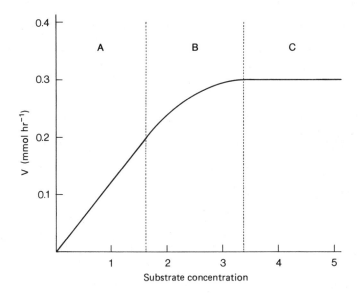

   (a) Which part of the curve (A, B, or C) represents a first-order reaction?
   (b) Which part represents a zero-order reaction?
   (c) What part of the curve could represent a substrate saturation of an enzyme?

References

Herman, W. A., McGill, W. B., and Dormaar, J. F. (1977). Effects of initial chemical composition on decomposition of roots of three different grass species. *Can. J. Soil Sci.* **57,** 205–215.

Jenkinson, D. S., and Rayner, J. H. (1977). The turnover of soil organic matter in some of the Rothamsted classical experiments. *Soil Sci.* **123**, 298–305.

Paul, E. A., and Van Veen, J. A. (1978). The use of tracers to determine the dynamic nature of organic matter. *Trans. Int. Congr. Soil Sci., 11th, 1978*, Vol. 3, pp. 61–102.

Paul, E. A., and Voroney, R. P. (1984). Field interpretations of microbial biomass and activity measurements. *In* "Current Perspectives in Microbial Ecology," (M. J. Klug and C. A. Reddy, eds.), pp. 509–514. ASM, Washington D.C.

Van Veen, J. A., Ladd, J. N., and Frissel, M. J. (1984). Modelling C & N turnover through the microbial biomass in soil. *Plant Soil* **76**, 257–274.

Voroney, R. P., Van Veen, J. A., and Paul, E. A. (1981). Organic C dynamics in grassland soils. 2. Model validation and simulation of long term effects of cultivation and rainfall erosion. *Can. J. Soil Sci.* **61**, 211–224.

Supplemental Reading

Clark, F. E., and Rosswall, T., eds. (1981). "Terrestrial Nitrogen Cycles," *Ecol. Bull.* (Stockholm), Vol. 23.

Haider, K., and Martin, J. P. (1975). Decomposition of specifically carbon-14 labelled benzoic and cinnamic acid derivates in soil. *Soil Sci. Soc. Am. Proc.* **39**, 657–662.

Legg, J. O., Chichester, F. W., Stanford, G., and DeMar, W. H. (1971). Incorporation of ^{15}N tagged mineral nitrogen into stable forms as soil organic nitrogen. *Soil Sci. Soc. Am. Proc.* **35**, 273–275.

Martin, J. P., and Haider, K. (1977). "Decomposition in Soil of Specifically [14]C Labelled PHP and Cornstalk Lignins, Model Humic Acid Type Polymers and Coniferyl Alcohols in Soil Organic Studies," Vol. II. IAEA, Vienna.

Piszkiewicz, D. (1977). "Kinetics of Chemical and Enzyme Catalyzed Reactions." Oxford Univ. Press, London, New York.

Sørensen, L. H., and Paul, E. A. (1971). Transformation of acetate carbon into carbohydrates and amino acid metabolites during decomposition in soil. *Soil Biol. Biochem.* **3**, 173–180.

Sowden, F. J., Chen, Y., and Schnitzer, M. (1977). The N distribution in soils formed under widely differing climatic conditions. *Geochim. Cosmochim. Acta* **41**, 1524–1526.

Warembourg, F. R., and Paul, E. A. (1977). Seasonal transfers of assimilated ^{14}C in grassland: Plant production and turnover, soil and plant respiration. *Soil Biol. Biochem.* **9**, 295–301.

Ziekus, J. G. (1982). Lignin metabolism and the carbon cycle. Polymer biosynthesis, biodegradation and environmental recalcitrance. *Adv. Microb. Ecol.* **5**, 211–243.

Chapter 8

Transformation of Nitrogen between the Organic and Inorganic Phase and to Nitrate

Introduction to the Nitrogen Cycle

Nitrogen is the mineral nutrient most in demand by plants and the fourth most common element in their composition, being outranked only by carbon, hydrogen, and oxygen. The cycling of nitrogen in nature has been studied more extensively than that of any other nutrient. The nitrogen atom exists in different oxidation states. Shifts between them are commonly mediated by soil organisms. Among the major nutrients, nitrogen has the greatest propensity to exist in the gaseous state. In the NO_3^- form, it is readily soluble in water and thus subject to leaching and water transport. In the NH_4^+ form it is subject to volatilization and to fixation both by clays and soil organic matter (SOM). The ease with which shifts occur in the oxidation states results in the formation of different inorganic forms that are readily lost from the ecosystem. Nitrogen shortages, therefore, often limit plant productivity. Also, both the gaseous and soluble phases of this nutrient lead to environmental pollution.

Following identification of the forms of nitrogen in soil and the role of microorganisms in moving nitrogen from one form to another, the concept of a nitrogen cycle was formulated by Löhnis (1913). He represented N_2 as central to the cycle and recognized the protein, amide, NH_3, NO_2^-, and NO_3^- forms. In following decades, the importance of such abiotic processes as leaching, volatilization, and decomposition became recognized, as did the effect of fertilizer nitrogen additions (Fig. 8.1). Since the 1950s, diagrams have been drawn to show both flow and compartmental values for different ecosystems. These tend to be very complex, and it is usually advantageous to analyze subcomponents of the cycle mathematically.

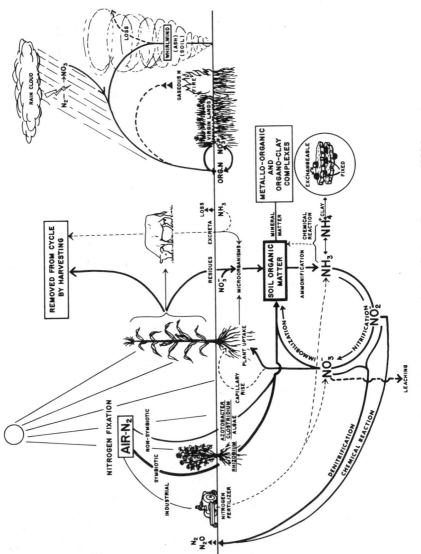

Figure 8.1. Nitrogen cycle in soil. (From Stevenson, 1982.)

Mineralization of Nitrogen

Three major biological forms of nitrogen are: proteins; microbial cell wall constituents, such as chitin and peptidoglycans; and the nucleic acids. Protein is a basic constituent of all life forms. During decomposition, it is hydrolyzed to peptides by proteinases and peptidases. The proteinases are classified as to whether they attack peptide linkages between specific amino acids. Examples include serine proteinases such as trypsin and subtilisin. Papain is an example of a sulfhydryl proteinase of plant origin, and pepsin is an acid protease. Acid proteases involve a metal ion, usually zinc, in the catalysis and have been purified from animal and fungal sources. The reaction mechanism is the reverse of that used in formation of peptide bonds. The nitrogen group receives a proton (H^+), and the carbon atom of the linkage receives an OH^- during the nucleophilic displacement reaction.

Mineralization of organic nitrogen refers to the degradation of proteins, amino sugars, and nucleic acids to NH_4^+, the mineral form. Whether an amino acid is used for an energy source or as a building block for other proteins is dependent on a complex series of feedback controls. Carbohydrate carbon, if available, will be utilized, rather than the amino acid carbon, thus preserving the carbon skeleton of the amino acid for protein synthesis. When deamination occurs, removal of NH_4^+ is most often carried out by enzymes such as glutamate dehydrogenase, which requires the coenzyme nicotine adenine dinucleotide (NAD) as acceptor of the reducing equivalents.

Whether NH_4^+ is immobilized or accumulates in the soil depends on the microorganism's requirement of nitrogen for growth. The carbon:nitrogen (C:N) ratio of microorganisms is not constant. Fungi can have wide C:N ratios (especially if phycomycetes, with cellulose in the cell walls, are present); their carbon contents are quite constant at approximately 45% carbon. With nitrogen contents of 3 to 10%, their C:N ratios range from 15:1 to 4.5:1. Bacteria have nitrogen in their cytoplasm and in the peptidoglycan of their cell walls. Their cellular nitrogen content is quite constant; C:N ratios usually are in the range 3:1 to 5:1.

Nonprotein sources of nitrogen include the cell wall constituents of bacteria and fungi. Amino sugars derived from bacterial cell walls or from fungal chitin constitute a major source of nitrogen during mineralization. Chitin degradation involves enzymes such as chitinase, which liberates the nitrogen acetylglucosamine monomers that constitute this polymer. The glucosamines are subsequently degraded by kinases that involve transfer of phosphate groups from ATP to form a glucosamine 6-phosphate. This is then deaminated to produce NH_4^+.

Nucleic acids, the constituents of DNA and RNA, are cyclic nitrogen compounds connected to phosphate groups by ester linkages. These are readily degraded in soil, and it is of interest that this group of compounds is one of the few constituents of living cells that does not appear to accumulate in SOM. The SOM, however, does contain up to 40% of its nitrogen in forms that are not recognized as coming from known biological sources. This nitrogen often is attached to phenolic compounds and is much more difficult to mineralize than that emanating from living organisms.

Immobilization Reactions

The nitrogen immobilization process involves the incorporation of NH_4^+ into amino acids. This depends on microbial growth, thus this process is closely tied to the availability of substrate carbon and abiotic parameters. The addition of NH_4^+ to α-ketoglutamic acid to form glutamic acid is one of the major reactions involved (Fig. 8.2). The reaction shown is reversible. The use of different coenzymes (NAD or NADP) by glutamate dehydrogenase during either the release or uptake of NH_4^+ allows organisms to control the courses of mineralization–immobilization processes. The glutamate dehydrogenase pathway incorporates NH_4^+ only at relatively high NH_4^+ levels. At levels more commonly found in soil, the microorganisms resort to another pathway named the glutamine synthetase–glutamine synthase pathway, which acts in two steps. This circuitous pathway (Fig. 8.3), in which two glutamates are formed from one glutamate, NH_4^+, and α-ketoglutarate, plays a major role in regulating the flow of NH_4^+ into amino acids. It is subject to cumulative feedback control by up to eight different inhibition reactions, which can act in an additive fashion to slow NH_4^+ incorporation.

Figure 8.2. Reductive amination pathway, which can be used to form glutamic acid.

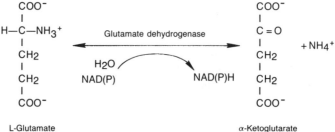

L-Glutamate   α-Ketoglutarate

Figure 8.3. Glutamine synthase–glutamate synthetase pathway for incorporation of NH_3 to form two glutamates. (From Mengel and Kirkby, 1978.)

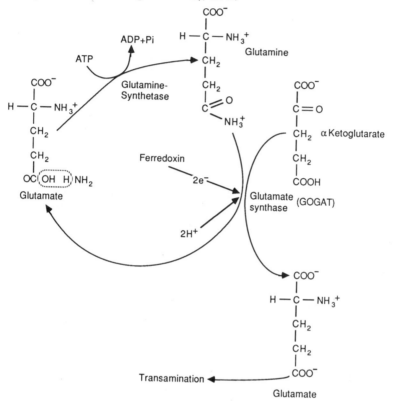

Once the nitrogen is incorporated into an organic form, further transfers to produce the numerous compounds within a cell involve transamination from glutamate to a keto acid. Figure 8.4 shows the transfer of the NH_4^+ group from glutamate to pyruvate to make the amino acid alanine. Each of the other amino acids requires the appropriate organic acid precursors. We have given examples only of the major pathways involved. Details for specific amino acids are available in biochemistry texts, such as that by Lehninger (1984).

Soil organisms generally have been shown to have a C:N ratio of 5:1 to 8:1. Since most microbial cells contain approximately 45% carbon, this represents a nitrogen content of 6 to 9%. The above C:N ratios are slightly lower than the average C:N ratio of soils (10:1), indicating that nonproteinaceous constituents also occur in SOM.

Figure 8.4.   Transfer of amino nitrogen in the formation of new amino acids.

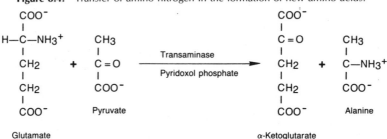

Knowledge of the growth efficiency of microorganisms on a simple substrate, such as glucose and mineral nitrogen, allows one to make calculations of the nitrogen requirements for growth. Figure 8.5 shows that 100 g of glucose, at 40% carbon, contains 40 g of carbon. A microbial yield coefficient of 60% results in 24 g of carbon in the biomass for a total

Figure 8.5.   Nitrogen requirements for growth of organisms with different nitrogen contents on glucose and inorganic nitrogen.

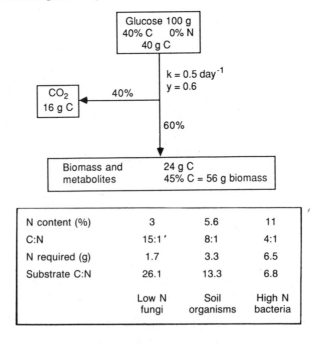

N content (%)	3	5.6	11
C:N	15:1 '	8:1	4:1
N required (g)	1.7	3.3	6.5
Substrate C:N	26.1	13.3	6.8
	Low N fungi	Soil organisms	High N bacteria

biomass weight of 56 g. Fungi, at a nitrogen content of 3%, would require a total of 1.7 g of nitrogen. Bacteria, with a nitrogen content of 11% (C:N ratio, 4:1), would immobilize 6.4 g of nitrogen. A mixed population, as commonly occurs in the soil, would require intermediate amounts of nitrogen.

Growth on a complex substrate with different decomposition rates, as shown in Chapter 7, requires a computer model to keep track of the various components and the formation of SOM. The incorporation of nitrogen flow into the model given for carbon in Fig. 7.5 yields the mineralization–immobilization model shown in Fig. 8.6 and Table 8.1.

Empirical research has demonstrated that decomposition of agricultural crop residues with 40% carbon and 1.6% nitrogen (C:N ratio, 25:1) will usually result in no net mineralization or immobilization. The net effect of the sum of both processes is zero, although both processes can be occurring at significant rates. If the plant C:N ratio is greater than 25:1, nitrogen will be taken up from the mineral nitrogen pool or degradation will be slowed until death of a part of the microbial population, or its simultaneous attack on SOM, which has a C:N of 10:1, releases more available nitrogen. These observations explain the generally accepted empirical formula that 1 kg of nitrogen should be added for each 100 kg of straw. This relationship holds because straw is often 0.6% nitrogen, whereas 1.6% nitrogen is required by the biomass for decomposition without immobilization of external nitrogen. Tree residues, with a high content of lignin that degrades slowly and is not substantially incorporated into microbial biomass, can show net mineralization at C:N ratios as high as 50:1. The fungi, which are the primary decomposers of forest litter, have the capability of attacking lignin to get at the associated nitrogen. They also transfer nitrogen from soil sources through their hyphae to the point of rapid growth. Thus decomposing litters often show increases in the amount of nitrogen present.

Once NH_4^+ has been formed there are a number of possible fates:

1. It can be taken up by plants and often is a preferred nitrogen source in solution. However, in most soils its positive charge leads to adsorption and decreased plant uptake.

2. It can be utilized for microbial growth. Many studies have indicated that NH_4^+, already being in the reduced state required for incorporation into amino acids, is preferred to NO_3^-. Even low levels of NH_4^+ often repress the enzymes required for NO_3^- reduction.

3. NH_4^+ is held on the exchange complex, where it can be replaced by cations in the soil solution.

Figure 8.6. Nitrogen mineralization–immobilization pools and fluxes in the degradation of plant residues and formation of soil organic matter. Flows: —— carbon; – – –, nitrogen. Numbers in circles refer to pools shown in Table 8.1.

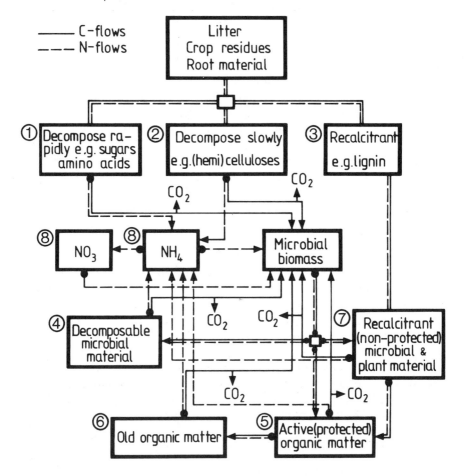

4. NH_4^+ is approximately the same size as the K^+ ion. It readily enters the interlayer portions of clays. The collapse of the interlayer space, for example, by drying, fixes the NH_4^+ so that it cannot be removed by exchange reactions, such as those occurring when extracting with a KCl solution. The amount of fixed NH_4^+ in the top meter of soil increases from 5 to 13% of the total nitrogen in surface horizons to 50% of the total nitrogen in subsurface horizons. The soil organic nitrogen content drops

Table 8.1

Pool Sizes, k Value, and Efficiency of Microbial Production Used in a Carbon–Nitrogen Turnover Model for the Decomposition of 1000 μgCg^{-1} of Straw in Soil

Pool	Residue carbon ($\mu g\ g^{-1}$)	Decomposition rate k (day^{-1})	Utilization efficiency (%)	Residue nitrogen ($\mu g\ g^{-1}$)	C:N ratio
Easily decomposable (1)[a]	150	0.2	60	25	6
Slowly decomposable (2)	650	0.08	40	12	54
Lignin and associated nitrogen (3)[b]	200	0.01	10[b]	3	67
Decomposable microbial products (4)[c]	6	0.8	40	1	6
Active protected soil organic matter (5)	5000	3×10^{-4}	20	555	9
Old organic matter (6)	7000	8×10^{-7}	20	700	10
Recalcitrant microbial and plant materials (7)	4	0.3	25	0.2	20
$NH_4^+ + NO_3^- - N$ (8)	—	—	—	—	—

[a]Numbers in parentheses refer to pools shown in Fig. 8.6.
[b]50% of lignin enters recalcitrant, nonprotected microbial and plant material pool.
[c]Actual pool sizes of biomass are much larger. These represent pool sizes produced from substrates in question. The soil preservation capacity to protect biomass would be applicable in a long-term field model.

drastically on cultivation or clear-cutting, but fixed NH_4^+ usually is retained at a higher percentage. It is only very slowly available to exchange reaction, and thus to microorganisms and plants.

5. NH_4^+ reacts with SOM to form quinone–NH_2 complexes. This is a significant reaction from a SOM-stabilization viewpoint.

6. The partial pressure of NH_3 is high. If present in the unadsorbed state, such as in senescing or decaying vegetation, in manure, or after application of urea in the presence of urease, high rates of NH_3 volatilization can occur. Surface manure in the field may lose up to 50% of its nitrogen due to volatilization.

7. NH_4^+ can be utilized as an energy source by a special group of autotrophs in the nitrification process.

Nitrification

Nitrification converts NH_4^+ to NO_2^- and NO_3^- nitrogen. The study of nitrification in soil is important from the standpoint of soil fertility, and because of the potential adverse impact that NO_3^- and its denitrification products can have on the environment. The phenomenon of NO_3^- formation was known in Europe as early as the fourteenth century, when saltpeter or niter (KNO_3) was first used as gunpowder and for pickling meat. In 1797, during the Napoleonic wars, the French survived the embargo of imported saltpeter, needed for the manufacture of gunpowder, by producing NO_3^- from niter heaps. Earth, manure, and lime were mixed in sheds and watered with urine and waste water. The heaps were kept aerated, and saltpeter was extracted with hot water. It was believed that NO_3^- was formed by a chemical reaction involving NH_4^+ and O_2, with the soil acting as a chemical catalyst.

In the 1870s, Pasteur postulated that the formation of NO_3^- was microbiological and analogous to the conversion of alcohol to vinegar. The first experimental evidence that nitrification was biological was provided by Schloesing and Müntz in 1877. They added sewage effluent to a long tube filled with sterile sand and $CaCO_3$. After 20 days, NH_4^+ had disappeared and NO_3^- was present. Heating of the column or addition of an antiseptic agent stopped the transformation; it could be reinitiated by addition of a small quantity of garden soil.

In 1878, Warington, at Rothamsted, England, found that nitrification was a two-stage process involving two groups of microorganisms. One microbial group oxidized NH_4^+ to NO_2^-, and another oxidized NO_2^- to NO_3^-. However, he did not isolate the responsible organisms. Winogradsky in 1890 was the first to isolate nitrifiers in pure culture.

Nitrifying Bacteria and Their Biochemical Reactions

The pioneering work of Winogradsky established that nitrification is typically associated with certain chemoautotrophic bacteria. They are obligate aerobes that derive their carbon solely from CO_2 or carbonates, and their energy from the oxidation of NH_4^+ or NO_2^-. The bacteria are classed into two groups, based on whether they oxidize NH_4^+ to NO_2^-, or NO_2^- to NO_3^- (Table 8.2). In most habitats the two are found together, and NO_2^- rarely accumulates in nature.

Figure 8.7 shows the microorganisms and the controls involved in the major inorganic nitrogen pathways. This shows that nitrification is part of the overall cycle that involves denitrification and nitrogen fixation as well as assimilatory and dissimilatory reduction. This figure also shows the possible pathways for heterotrophic nitrification via hydroxymates and primary nitro compounds to generate NO_2^- and NO_3^- but not free energy. Payne's diagram shows the oxidation of ammonia (NH_3) to nitrite (NO_2^-) by the genera *Nitrosoccus* and *Nitrobacter*. This proceeds via the intermediate hydroxylamine (NH_2OH) (Payne, 1981).

The oxidation of NH_4^+ to hydroxylamine (NH_2OH) is an endothermic reaction. The next reactant shown has not been identified, but it is hy-

Table 8.2
Listing of Chemoautotrophic Nitrogen Oxidizers[a]

Genus	Species	Habitat
Oxidize ammonia (NH_3) to nitrite (NO_2^-)		
Nitrosomonas	europaea	Soil, water, sewage
Nitrosospira	briensis	Soil
Nitrosococcus	nitrosus	Marine
	oceanus	Marine
	mobilis	Soil
Nitrosovibrio	tenuis	Soil
Oxidize nitrite (NO_2^-) to nitrate (NO_3^-)		
Nitrobacter	winogradskyi[b]	Soil
	(agilis)[b]	Soil, water
Nitrospira	gracilis	Marine
Nitrococcus	mobilis	Marine

[a]From "Bergey's Manual of Determinative Bacteriology," 8th edition.
[b]Nitrobacter winogradskyi is comprised of at least two serotypes, one of which, prior to the 8th edition, was referred to as N. agilis.

pothesized that it is a nitroxyl radical; NO_3^- is thought to react with this radical to form nitrohydroxylamine, which then breaks down to two HNO_2 molecules. Oxidation of NO_2^- to NO_3^- by *Nitrobacter* also involves O_2, but its role is confined to electron transport. The oxygen atom in NO_3^- is generated from H_2O, not from O_2. The overall reactions produce energy for growth. The resulting free energy change ΔF is -65 and -18.2 kcal mol^{-1} for the oxidation of a mole of NH_4^+ and NO_2^-, respectively.

Factors Affecting Nitrification

The soil perfusion apparatus has traditionally been used for the study of the factors affecting nitrification. In its operation, a metabolite is continuously percolated through a soil column. The repeated perfusion permits direct study of kinetics of the transformation and the effects of environmental change. Soil columns receiving water and substrate via multichannel pumps and interfaced with computer-controlled sampling devices are now also available such that the soil process can be measured on a continuous basis.

Acidity

A significant correlation between NO_3^- production and pH has been demonstrated. Optimum pH values may vary between 6.6 to 8.0. Typically, nitrification rates in agricultural soils decrease markedly below pH 6.0

THE NITROGEN CYCLE : A MICROBIOLOGICAL PERSPECTIVE

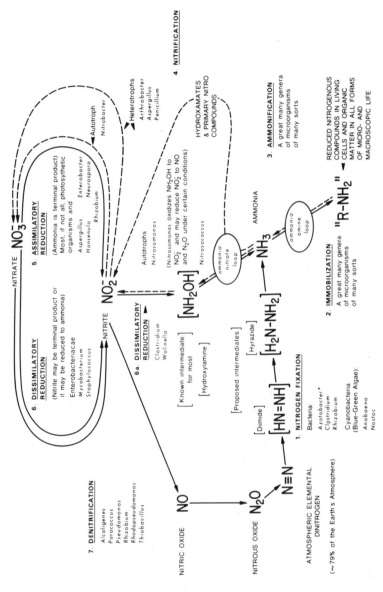

* Listings include as examples only a few genera containing active species for each numbered phenomenon; ——► REDUCTIVE STEPS; --—► OXIDATIVE STEPS

W. J. Payne 1986

Figure 8.7. Nitrification reactions in the nitrogen cycle. (Courtesy, W. J. Payne, personal communication.)

and become negligible below 4.5. At high pH, NH_4^+ inhibits the transformation of NO_2^- to NO_3^-. Acidic forest soils nearly all produce NO_3^- when disturbance, such as fire or clear-cutting, results in higher rates of nitrogen mineralization than immobilization. That heterotrophic nitrifiers and/or microsites may account for the fact that nitrification in forest soils is not as acid sensitive as that in agricultural soils was noted in Chapter 2.

Aeration

Since O_2 is an obligate requirement for all species concerned, aeration is essential for nitrification. Diffusion of O_2 into soil, and therefore, aeration, is controlled by factors such as soil moisture and soil structure.

Moisture

Because moisture affects the aeration regime of the soil, the water status of the soil has an influence on NO_3^- production. Waterlogging limits diffusion of O_2, and nitrification is suppressed. On the other extreme, bacterial proliferation is retarded by an insufficiency of water. The optimum moisture level varies with different soils, but in most, nitrification proceeds readily at -0.1 to -1 MPa moisture tension. General mineralization reactions producing NH_4^+ are less sensitive to both water stress and low temperature; NH_4^+ therefore accumulates in water-stressed or cool soils.

Temperature

Nitrification is markedly affected by temperature. The process is slow below 5°C and above 40°C. The optimum is around 30 to 35°C. The interaction of temperature, moisture, aeration, and other factors makes up the seasonal effect. In temperate areas, nitrification is greatest in spring and fall and slowest in summer and winter.

Organic Matter

For a long time it was thought that organic matter was toxic to nitrifiers. This could not be accepted because the process occurs in manure piles and sewage. It is now believed that organic matter per se is not inhibitory, but its decomposition may require inorganic nitrogen and O_2, thus depleting supplies of available NH_4^+ and O_2 for the nitrifiers.

Heterotrophic Nitrification

Several heterotrophic bacteria and actinomycetes are able to generate traces of NO_2^- when grown in culture media containing NH_4^+. A few bacteria, such as strains of *Arthrobacter,* and fungi, such as *Aspergillus flavus,* can produce NO_3^- from NH_4^+. The heterotrophs do not derive energy from these oxidations, and their significance in nature is not known. The production of NO_3^- in agricultural soils by heterotrophs appears to be insignificant in relation to that brought about by the chemoautotrophs.

Undesirable Effects of Nitrification

Because plants readily assimilate NO_3^-, biological nitrification has been used as an index of soil fertility, and good plant growth has often been considered dependent on this process. However, nitrification can also lead to undesirable consequences. Ammonium is cationic, adsorbs to soil, and is relatively stationary. Nitrate, on the other hand, is an anion and is freely mobile in soil solution. Under certain conditions, particularly in sandy soils, under heavy rainfall, or where excessive irrigation is practiced, NO_3^- will leach away from the root zone. It is also susceptible to losses through denitrification. This results in atmospheric contamination. Excess NO_3^- nitrogen leached from soil often ends up in lakes and streams, where it has been implicated in (1) excess growth of plants and algae (eutrophication), (2) health problems, such as infant and animal methemoglobinemia, and (3) formation of carcinogenic nitrosamines by reaction with other nitrogenous compounds.

Inhibition of Nitrification

Nitrate usually does not accumulate in undisturbed grasslands or mature forests. There have been numerous suggestions that climax ecosystems produce organic compounds, such as tannins, that are toxic to nitrifiers. This process, called allelopathy, would represent a natural control of nitrogen losses. The alternate argument, that competition for NH_4^+ by plant roots, mycorrhizal fungi, and nitrogen-immobilizing microorganisms keeps inorganic nitrogen rates low, is now more widely accepted.

To reduce losses of nitrogen, a great deal of research has been devoted to finding a compound that will inhibit the nitrification process. While several have been patented, no inhibitor has been adopted for general use. Table 8.3 lists some of the patented inhibitors now available. Factors such as soil type, temperature, pH, and moisture affect each chemical differently. Another approach is to develop slow-release fertilizers. Here the objective is to create fertilizer materials that will release only small quantities of inorganic nitrogen at a time, thus improving fertilizer efficiency by avoiding NO_3^- build-up and losses. Sulfur-coated urea is one such compound. Because there is generation of acidity during nitrification, the

Table 8.3
Some Patented Nitrification Inhibitors[a]

Chemical	Common name	Developer	Inhibition (% by day 14)
2-Chloro-6-(trichloromethyl)pyridine	N-serve	Dow Chemical	82
4-Amino-1,2,4-6-triazole-HC1	ATC	Ishihada Industries	78
2,4-Diamino-6-trichloromethyltriazine	CL-1580	American Cyanamid	65
Dicyandiamide	Dicyan	Showa Denko	53
Thiourea	TU	Nitto Ryuso	41
3-Mercapto-1,2,4-triazole	MT	Nippon	32
2-Amino-4-chloro-6-methylpyrimidine	AM	Mitsui Toatsu	31
Sulfathiazole	ST	Mitsui Toatsu	31

[a]From Bundy and Bremner (1973).

avoidance of excess NO_3^- accumulation is an important practical consideration.

References

Alexander, M. (1977). "Introduction to Soil Microbiology." Wiley, New York.

Bundy, L. G., and Bremner, J. M. (1973). Inhibition of nitrification in soils. *Soil Sci. Soc. Am. Proc.* **37**, 396–398.

Campbell, R. (1977). Microbial ecology. *In* "Basic Microbiology" (J. F. Wilkinson, ed.), Vol. 5. Wiley, Toronto.

Lehninger, A. L. (1984). "Biochemistry." Worth Publishers, New York.

Lohnis, F. (1913). "Vorlesungen über landwortschaftliche Bacteriologia." Borntraeger, Berlin.

Mengel, K., and Kirkby, E. A. (1978). "Principles of Plant Nutrition." Int. Potash Inst., Bern.

Miller, H. G., Cooper, J. M., Miller, J. D., and Pauline, D. J. (1979). Nutrient cycles in pine and their adaptation to poor soils. *Can. J. For. Res.* **9**, 19–26.

Nommik, H., and Vahtras, K. (1982). Retention and fixation of ammonium and ammonia in soils. *In* "Nitrogen in Agricultural Soils" (F. J. Stevenson, ed.), Agronomy, Vol. 22, Am. Soc. Agron., Madison, Wisconsin.

Payne, W. J. (1981). "Denitrification." Wiley, New York.

Schloesing, J. J. T., and Müntz, A. (1877). Sur la Nitrification par les Ferments Organisés. *Comp. Rend Acad. Sci (Paris)* **84**, 301–303.

Stevenson, F. J., ed. (1982). "Nitrogen in Agricultural Soils," Agronomy, Vol. 22. Am. Soc. Agron., Madison, Wisconsin.

Warington, R. Nitrification. *J. Chem. Soc.* **33**, 44–51.

Winogradsky, S. (1890). Recherches sur les organisms de la nitrification. *Ann. Inst. Pasteur* **4**, 213–231, 257–275, 760–771.

Supplemental Reading

Belser, L. W. (1982). Inhibition of nitrification. *In* "Advances in Agricultural Microbiology" (N. S. Subba Rao, ed.), pp. 267–293. Butterworth, London.

Focht, D. D., and Verstraete, W. (1977). Biochemical ecology of nitrification and denitrification. *Adv. Microb. Ecol.* **1,** 135–199.

Gosz, J. R. (1981). Nitrogen cycling in coniferous ecosystems. *Ecol. Bull.* (Stockholm) **33,** 405–426.

McGill, W. B., Hunt, H. W., Woodmansee, R. G., and Reuss, J. O. (1981). Phoenix, a model of the dynamics of carbon and nitrogen in grassland soils. *Ecol. Bull.* (Stockholm) **33,** 49–115.

Schmidt, E. L. (1978). "Nitrifying Microorganisms and Their Methodology," pp. 288–291. Am. Soc. Microbiol., Washington, D.C.

Schmidt, E. L. (1982). Nitrification in soil. *In* "Nitrogen in Agricultural Soils" (F. J. Stevenson, ed.), Agronomy, vol. 22, pp. 253–288. Am. Soc. Agron., Madison, Wisconsin.

Van Veen, J. A., Ladd, J. N., and Frissel, M. J. (1984). Modelling C & N turnover through the microbial biomass in soil. *Plant Soil* **76,** 257–274.

Chapter 9

Reduction and Transport of Nitrate

Introduction

In most soils, NH_4^+ released during soil organic matter (SOM) decomposition and not immediately reused by organisms is rapidly transformed to NO_3^-. Fertilizer nitrogen, added as urea, NH_3, or in the NH_4^+ form, is also subject to nitrification. Once NO_3^- is formed in soil, it is subject to the following fates: (1) it may undergo denitrification by microorganisms to gaseous oxides of nitrogen and to N_2, (2) it may be taken up by organisms and used in synthesis of amino acids (assimilatory reduction), (3) in the absence of O_2 it may be used by microorganisms as an electron acceptor and become reduced to NH_4^+ (dissimilatory reduction), (4) it may be leached from the upper profile to deeper soil layers or to ground water, (5) it may be transported off site by runoff, or (6) it may accumulate in the soil, as it often does under fallow conditions.

Denitrification

Microbial reduction of NO_3^- to NO_2^- and then to gaseous N_2O and N_2, which are commonly lost to the atmosphere, is known as denitrification. This process, often called enzymatic denitrification, must be distinguished from the assimilatory reduction of NO_3^- accomplished by various organisms during growth and also from the dissimilatory reduction of nitrate to ammonium that is accomplished by certain microorganisms in the absence of O_2 (see Fig. 8.7). In assimilatory reduction, green plants, bacteria, cyanobacteria, and fungi reduce NO_3^- to NH_4^+ in the course of biosynthesis of amino acids and protein. Nitrite and hydroxylamine

(NH$_2$OH) are intermediate products; the process is not inhibited by O$_2$ but is repressed by the presence of NH$_4^+$ or reduced nitrogenous organic metabolites. The reductase is linked to NADPH; molybdenum is a required cofactor. In dissimilatory reduction, certain microbial species use NO$_3^-$ as an electron acceptor in the absence of O$_2$ to produce NH$_4^+$, but not N$_2$, as an end product. The likely pathway for this reaction is as follows:

$$2 \text{ HNO}_3 \xrightarrow[- 2 \text{ H}_2\text{O}]{+ 4 \text{ H}} 2 \text{ HNO}_2 \rightarrow ? \rightarrow [\, 2 \text{ NH}_2\text{OH}] \xrightarrow[-2 \text{ H}_2\text{O}]{+ 6 \text{ H}} 2 \text{ NH}_4^+$$

This pathway does not lead to nitrogen loss from the system. It occurs under highly reducing conditions, but the extent of its occurrence in nature is not well defined. In soil it is difficult, unless tracers such as ^{15}NO$_3^-$ are utilized, to differentiate this pathway from proteolysis, which also yields NH$_4^+$. An extensive number of microorganisms can reduce NO$_3^-$ only to NO$_2^-$ in the presence of organic matter and the absence of O$_2$. This process, called nitrate respiration, does not lead to direct losses, but NO$_2^-$ is very reactive and can indirectly lead to losses.

Biochemistry of Denitrification

The sequence of identifiable products formed during denitrification as shown in Fig. 9.1 is nitrate (NO$_3^-$), nitrite (NO$_2^-$), nitrous oxide (N$_2$O) and nitrogen gas (N$_2$). Under field conditions, not all of the intermediate gaseous products are converted to N$_2$ by each of the specific reductase enzymes involved; portions of them escape to the atmosphere. The comprehensive view of electron flow during denitrification (Fig. 9.2) shows that a variety of electron donors can be utilized in the electron transport mechanism involved in denitrification. The quinone component of the electron transport scheme is shown as the one probably influenced by the presence of O$_2$; when O$_2$ is not available as an electron acceptor, electron transport branches off from b-type cytochromes to the several forms of oxidized nitrogen as the terminal acceptor. Specific reductases are involved at each acceptor level. Nitrate (NO$_3^-$) reductase contains molybdenum, iron, and labile sulfide groups. Both molybdenum and iron are necessary constituents for enzyme activity.

The nitrite (NO$_2^-$) reductase functioning at the second step of denitrification is easily solubilized by cell manipulation and may at different times be found in the soluble fraction of the cytoplasm but is thought to be membrane bound. As with nitrate (NO$_3^-$) reductase, K_m values are wide ranging and may reflect some dependence on anion transport within the cell or inhibition by reaction products.

Figure 9.1. Products formed during denitrification in Melville loam, pH 7.8. (From Cooper and Smith, 1963.)

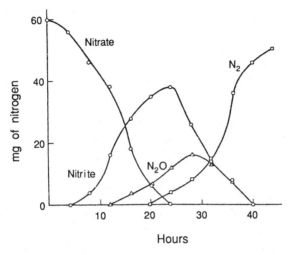

The membrane bound nitric oxide (NO) reductase, has been isolated but not yet purified; its characteristics are therefore not well described (Table 9.1). W. Payne's scheme in Fig. 9.2 shows NO as an intermediate. There is, however, some controversy as to the pathway by which nitrite (NO_2^-) is reduced to nitrous oxide (N_2O). One hypothesis states that nitric oxide (NO) is formed from the reduction of the nitrosyl to a free nitroxyl (HNO) which spontaneously dimerizes to N_2O. Work in the laboratory of J. Tiedje argues for the nitric oxide (N_2O) being produced by nucleophilic attack of nitrite on an enzyme-bound nitrosyl intermediate rather than via free nitroxyl. The scheme for the formation of the N—N bond by nucleophilic attack of nitrite on an enzyme-bound nitrosyl (Aerssens *et al.*, 1986) is as follows:

$$E-Fe^{II} + NO_2^- \rightleftharpoons E-Fe^{II} \cdot NO_2^- \overset{\pm H_2O}{\rightleftharpoons} E-Fe^{II} \cdot NO^+$$

$$\Big\| \pm NO_2^-$$

$$N_2O + E-Fe^{II} \leftarrow \leftarrow E-Fe^{II} \cdot (N_2O_3)$$

$$(E = \text{Enzyme}, Fe^{II} = \text{Iron Group})$$

The nitrous oxide (N_2O) reductase is unique among the several reductases in that it is inhibited by acetylene (C_2H_2) and sulfide (HS^-). This susceptibility has provided a convenient method for use in studies of denitrification. Normally, N_2O and N_2 are produced in varying ratios,

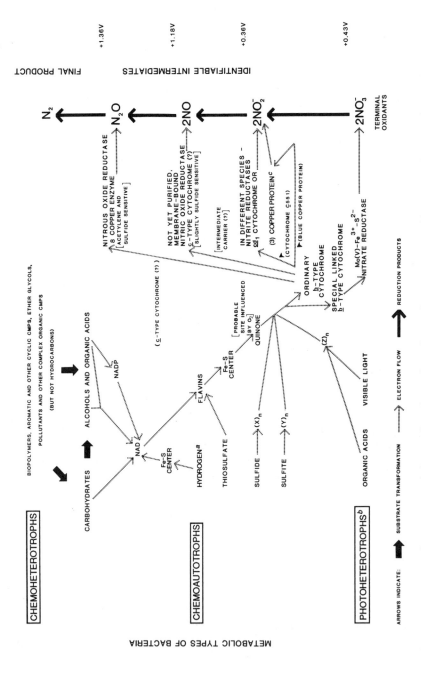

Figure 9.2. A comprehensive view of electron flow during denitrification by various types of bacteria under anaerobic conditions (courtesy W. J. Payne, personal communication)

Table 9.1
Properties of Reductases Involved in Denitrification[a]

Property	$NO_3^- \rightarrow$ NaR[b]	$NO_2^- \rightarrow$ NiR[c]	$NO \rightarrow$ NOR[d]	$N_2O \rightarrow N_2$ N$_2$OR[e]
Location				
Soluble	−	+	+	−
Membrane	+	+?		+
Periplasmic		+		
Phophorylation occurs	+	+?	?	+
Oxygen				
Represses	+	+	+	+ +
Inhibits	+	+?	+?	+ +
NH_4^+ represses	−	−	−	−
Nitrogen oxide required as inducer	−	−	−	−
NO_3^- inhibits	−	−	+	+
Low pH inhibits	+	+	+	+ +
Sulfide inhibits	±	±	?	+ + +
C_2H_2 inhibits	−	−	−	+ + +

[a]From Knowles (1981).
[b]Nitrate (NO_3^-) reductase.
[c]Nitrite (NO_2^-) reductase.
[d]Nitric oxide (No) reductase.
[e]Nitrous oxide ($N_2$0) reductase.

depending on the substrate, environmental conditions, the organisms involved, and also on the time elapsed since the onset of denitrifying activity. With C_2H_2 blockage, N_2O accumulates and can be measured by gas chromatography. That the C_2H_2 does not inhibit the onset and progress of the denitrification prior to the N_2O blockage has been shown in experiments using tracer nitrogen. Criticism of the $C_2\dot{H}_2$ blockage technique has been made on various grounds. There may exist, in soil, unknown denitrifiers whose reductase is not inhibited over long periods of time. Adequate diffusion of C_2H_2 to active microsites of denitrification may not be accomplished during short periods of time. However, field results with nitrogen-15 have generally substantiated the reliability of the C_2H_2 blockage technique and it is a rapid, sensitive assay.

Chemodenitrification

Nitrite reductases are not solely responsible for the dismutation of NO_2^- in soil. It is also subject to chemical reactions that lead to production of N_2 by nonenzymatic pathways under fully aerobic conditions. Losses of

NO_2^- nitrogen because of its instability and dismutation to other oxides of nitrogen and to N_2 are commonly referred to as chemodenitrification. At times such losses have been termed aerobic denitrification, but this term is undesirable in that it has also been used for the microbial denitrification occurring within anaerobic microsites in apparently well-aerated soils. Chemodenitrification is also in part a misnomer, as the production of NO_2^- in soil is a biotic process.

Aerobic dismutation of nitrite can occur by several pathways. In the Van Slyke reaction, amino groups in the α position to carboxyls yield N_2, as follows:

$$RNH_2 + HNO_2 \rightarrow ROH + H_2O + N_2$$

In soil in the presence of air, this reaction occurs at an appreciable rate only at values of pH 5 or lower.

In reactions quite similar to the Van Slyke reaction, NO_2^- reacts with ammonia, urea, methylamine, purines, and pyrimidines as follows:

$$HNO_2 + NH_4^+ \rightarrow N_2 + 2 H_2O$$

Under comparable conditions this reaction proceeds more slowly than the Van Slyke reaction. Chemical decomposition of nitrous acid may occur spontaneously, as follows:

$$3 HNO_2 \rightarrow HNO_3 + H_2O + 2 NO$$

The NO may escape to the air or react with O_2 and water to form NO_3^-:

$$2 NO + O_2 \rightarrow 2 NO_2 \rightarrow N_2O_4 + H_2O \rightarrow HNO_2 + HNO_3$$

Several workers have suggested that NH_4NO_2 instability may be involved in chemodenitrification. It has been noted that both NH_4^+ and NO_2^- are often simultaneously present in soil; there could be N_2 loss associated with the formation and decomposition of NH_4NO_2:

$$NO_2^- + NH_4^+ \rightarrow NH_4NO_2 \rightarrow N_2 + H_2O$$

Clark (1962) and coworkers noted that although NO_3^- added to well-aerated soils could be quantitatively recovered after several weeks of incubation, NH_4^+ or urea nitrogen added in parallel incubations was recoverable in some soils but not in others. Deficits of up to one-half the initially added nitrogen were encountered in soils showing high NO_2^- accumulation during the nitrification process. The investigators suggested that organic reducing compounds generated by heterotrophic organisms reacted with NO_2^- to cause its chemical decomposition in neutral and weakly alkaline soils as well as in acidic soils.

Organisms Capable of Denitrification

A large number of bacteria can reduce NO_3^- to NO_2^- in the absence of O_2. A much smaller number of species, for the most part occurring in the genera listed in Table 9.2, can carry the reduction to N_2O and N_2. Most denitrifiers are heterotrophic and are members of commonly occurring soil genera, such as *Pseudomonas, Bacillus,* and *Alcaligenes.* Most are aerobic organisms that grow anaerobically only in the presence of nitrogen oxides. However, some species of *Propionibacterium* are obligately fermentative, and some species of *Bacillus* are facultatively fermentative. Most denitrifers possess NO_3^-, NO_2^-, and N_2O reductases, but there are some that lack NO_3^- reductase, and others (principally *Pseudomonas*) that lack N_2O reductase.

Both denitrifying and nitrogen-fixing capabilities exist in some members of *Azospirillum* and *Rhizobium,* but the two processes do not occur simultaneously, and the presence of NO_3^- represses nitrogenase activity. Certain photosynthetic nonsulfur bacteria of the genus *Rhodopseudomonas* also have the interesting capability for both denitrification and N_2 fixation. As well, there are lithotrophic denitrifiers. *Thiobacillus denitrificans* is a sulfur oxidizer that under anaerobic conditions carries out the following reaction:

$$5 \text{ S} + 6 \text{ KNO}_3 + 2 \text{ H}_2\text{O} \rightarrow 3 \text{ N}_2 + \text{K}_2\text{SO}_4 + 4 \text{ KHSO}_4$$

Table 9.2
Genera of Bacteria Capable of Denitrification[a]

Genus	Interesting characteristics of some species
Alcaligenes	Commonly isolated from soils
Agrobacterium	Some species plant pathogens
Azospirillum	Capable of N_2 fixation, commonly associated with grasses
Bacillus	Thermophilic denitrifiers reported
Flavobacterium	Denitrifying species isolated
Halobacterium	Requires high salt concentrations for growth
Hyphomicrobium	Grows on one-carbon substrates
Paracoccus	Capable of both lithotrophic and heterotrophic growth
Propionibacterium	Fermentors capable of denitrification
Pseudomonas	Commonly isolated from soils
Rhizobium	Capable of N_2 fixation in symbiosis with legumes
Rhodopseudomonas	Photosynthetic
Thiobacillus	Generally grow as chemoautotrophs

[a]From Firestone (1982).

It also oxidizes partially reduced sulfur compounds, as follows:

$$5 \ K_2S_2O_3 + 8 \ KNO_3 + H_2O \rightarrow 4 \ N_2 + 9 \ K_2SO_4 + H_2SO_4$$

Procedures for Measuring Denitrification

Determining the extent of denitrification in different soils, sediments, and aquatic systems has been hampered by the lack of fully adequate methodology. Some of the approaches that have been used are listed in Table 9.3. Nitrogen balance studies using nitrogen-15 provide indirect but valuable estimates, especially when leaching losses can be accounted for, as in lysimeters or monitored watersheds. Given the great number of variables and uncertainties that exist in the field, nitrogen-15 has proven useful, but difficulties arising in the spectrometry of mixtures of N_2, N_2O, NO, and NO_2 cloud the determination of constituents contributing to specific mass

Table 9.3
Methods for Measurement of Denitrification in Soils[a]

Method	Comments
Nitrogen or nitrogen-15 balance	Must be relatively long term; low precision requires addition of nitrogen or nitrogen-15
Disappearance of NO_3^- or NO_2^-	Valid in some conditions; may be assimilated or reduced to NH_4^+, especially when carbon rich and highly anaerobic
Disappearance of added N_2O	Nondisturbing; saturation of N_2O-reducing sites may overestimate rate; NO_3^- or or NO_2^- may compete for electrons
Production of N_2O and (or) N_2	Requires closed system with air N_2 replaced by argon or helium; changes O_2 concentrations
Production of $^{15}N_2O$ and (or) $^{15}N_2$ from $^{15}NO_3^-$	
In laboratory systems	Commonly used in closed systems; precision relatively high but tedious and expensive
In the field	Fluxes of nitrogen-15 gases determined in short-term enclosures; insensitive
Reduction of $^{13}NO_3^-$ to $^{13}N_2O$ and (or) $^{13}N_2$	High sensitivity but very short (10 min) half-life; used mainly in stirred slurries to promote mixing
Production of N_2O in the presence and absence of C_2H_2	N_2O reductase is inhibited by C_2H_2, and N_2O accumulates
In laboratory systems	Appears to be reliable in 0.1 atm C_2H_2
In the field	Flow-through or gas-tight enclosures; incomplete inhibition sometimes observed; C_2H_2 may be utilized, or it may inhibit nitrification and other processes

[a]From Knowles (1982).

peaks. For example, both $^{15}N_2$ and NO contribute to mass peak 30 in mass spectrometry.

The availability of good detectors and packing materials has made gas chromatography a valuable tool, especially in systems in which N_2 is absent in the initial gas phase. At normal atmospheric N_2 concentrations, present technology cannot detect small changes in N_2. Therefore, measurement of N_2O in the soil profile is often used as an index of denitrification. The actual amount occurring is difficult to calculate because of uncertainty in the $N_2:N_2O$ ratio and differences in their diffusion rates. Moreover, using the rate of N_2O reduction as an index of denitrification can be criticized on the ground that N_2O is usually not the rate-limiting step in denitrification.

Following demonstrations that C_2H_2 blocked reduction of N_2O to N_2 by soil bacteria, such blockage has been extensively exploited in studies of denitrification. The blockage has been shown to be reversible at C_2H_2 concentrations below 0.02 atm; at higher concentrations the blockage is not reversible, and there is negligible subsequent reduction of any N_2O produced. The accumulation of N_2O can readily be measured by a gas chromatograph fitted with an electron capture detector. The gaseous sample must be taken from a soil column enclosed in a container or under a canopy. The extrapolation of these measurements to take spatial heterogeneity and yearly effects into account requires a good knowledge of the soil variability and the climate changes involved.

Soil Factors Affecting Denitrification

For denitrification to occur, oxidized nitrogen must be present, O_2 availability must be limited, and denitrifying organisms must be present in an environment favorable for their growth. Important determinants of the rate of denitrification in soil are (1) availability of NO_3^-, (2) the supply of metabolizable carbon, (3) soil moisture, (4) soil aeration, (5) soil pH, and (6) temperature. The qualitative roles of these determinants taken separately are fairly well defined, but their interactions make difficult the prediction of actual fluxes of gaseous nitrogen from field soils. Mosier *et al.* (1982) have offered a simple mechanistic model for prediction of daily N_2O losses from soil.

Soil Nitrate Content

Inasmuch as denitrification is enzyme mediated, substrate concentration should function as a rate determinant. At NO_3^- concentrations exceeding 20 μg nitrogen ml^{-1}, the denitrification reaction follows zero-order kinetics; e.g., it is independent of the amount of NO_3^- present. The reaction rates of NO_3^- concentrations may be determined by carbon availability for metabolism rather than by NO_3^- level. At low NO_3^- nitrogen concentrations, the kinetics of reduction appear to be first order. At low levels in wet soil, the determinant also may be the rate of NO_3^- diffusion to the site of denitrification. Nitrate concentration has been observed to influence the $N_2:N_2O$ ratio in the gaseous products of denitrification. At high NO_3^- levels N_2 is predominant, and at low levels it is often N_2O.

Carbon Availability

Most denitrification is accomplished by heterotrophic bacteria, and therefore, the process is strongly dependent on carbon availability. There is general correlation between total SOM content and denitrification potential, but much better correlation occurs with the supply of easily decomposable organic matter. Available carbonaceous substrates supply electrons. Their decomposition also produces CO_2 and reduces O_2 levels, thus increasing the demand for NO_3^- as an acceptor of electrons during microbial growth. Attempts have been made to measure readily decomposable organic matter to predict soil denitrifying capacities. The amount of water-soluble carbon measurable in soil has been found to account for 71% of its denitrification potential, and the amount of carbon mineralizable during 7 days of incubation was more than adequate to account for all of the requirements.

The increased amount of easily decomposable carbon provided by root exudates and root exfoliates in the rhizosphere suggests that this region may be more conducive to denitrification than nonrhizosphere soil. With incubation conditions held constant, laboratory experiments with rhizosphere soil and neighboring soil only a few millimeters distant have shown higher denitrification in the former. Field experiments have been contradictive. Instances of decreased denitrification potentials in cropped versus fallow soils are attributable to NO_3^- and water uptake by the plant. Evapotranspiration provides a better-aerated soil in the vicinity of plant roots. The length of time that the rhizosphere is wet is another variable. During short-term irrigations or ponding of water, the additional available carbon in the rhizosphere may be influential, but with longer-term flooding and the establishment of a low redox potential, the rhizosphere effect becomes inconsequential.

Soil Water Content

The influence of water is linked to its role in governing O_2 diffusion to sites of microbial activity. Increases in soil water content to levels that interfere with air diffusion progressively increase denitrification potential. Oxygen is repressive to all the nitrogen oxide reductases, but slightly less

Table 9.4

Total Denitrification (μmol N_2O g^{-1} in the Presence of 0.1 atm C_2H_2) and (in Parentheses) the Average mole fraction[a] of N_2O in the Products during 3 Days of Incubation of Soil[b] at Different Water and O_2 Concentrations[c]

Water content (% MWHC[d])	Initial percentage of O_2 in the gas phase							
	0		4		8		12	
33	0.64	(0.94)	0	—	0	—	0	—
60	0.98	(0.72)	0.03	(3.6)	0	—	0	—
100	0.98	(0.83)	0.68	(0.69)	0.36	(0.25)	0.34	(0)
200	1.04	(0.97)	1.09	(0.43)	0.90	(0.21)	0.58	(0)

[a]The mole fraction of N_2O is the N_2O produced in the absence of C_2H_2, divided by the N_2O produced in the presence of 0.1 atm C_2H_2.
[b]For incubation details, see footnote [c].
[c](From Knowles, 1981.)
[d]MWHC, maximum water-holding capacity.

repressive to NO_3^- reductase than to the others. The interrelationships of water and O_2 in soil are complicated by the existence of soil aggregates, soil pores, and channels of diverse sizes and shapes. Wet soils are often cold soils, thereby reducing biochemical oxygen demand. Table 9.4 shows that for a given soil water content, denitrification decreases with increasing O_2 content; it increases with increasing water content. Generally, in the field, and providing there is not an abnormally high O_2 consumption rate, denitrification is lacking or insignificant at moisture levels less than 60% of moisture holding capacity, which is approximately equivalent to −0.01 MPa moisture stress.

Soil pH and Temperature

Most denitrifying bacteria grow best near neutrality (pH 6–8). Denitrification becomes slow but may still remain significant below pH 5 and is negligible or absent below pH 4. Degree of soil acidity also influences the $N_2O:N_2$ ratio in the evolved gases.

Soil temperature affects denitrification directly in that microbial activity increases exponentially with increasing temperature according to the Arrhenius equation, and indirectly in that temperature affects both O_2 solubility and O_2 diffusion in water. In cold soils (less than 15 to 20°C) a linear rather than exponential relationship exists between temperature and denitrification potential.

The minimum temperature for denitrification is about 5°C, and maximum temperature, about 75°C. At soil temperatures near the upper limit, thermophilic bacilli are primarily but not necessarily wholly responsible for gaseous loss; above 50°C, chemical decomposition of NO_2^- may become significant and supplement enzymatic decomposition. In some, but not all, soils, temperature changes have been noted to affect the $N_2O:N_2$ ratio in the evolved gases.

Denitrification Losses Relative to Other Fates of Soil Nitrogen

Nitrogen balance studies, have frequently been used to obtain an indirect estimate of denitrification. Estimates such as those in Table 9.5 show an average of 15% of fertilizer nitrogen being lost to the atmosphere.

With the advent of C_2H_2 blockage methodology for measuring denitrification, measurement of gaseous losses have been made for a variety of soils and cropping conditions. Fluxes of N_2O from soil are not solely indicative of denitrification. In soil untreated with C_2H_2, N_2O is produced during nitrification. In soil fertilized with nitrifiable nitrogen (NH_4^+, urea), the N_2O flux ascribable to nitrification can be significant part of the total nitrogen oxide flux.

Impact of Nitrous Oxide on the Stratosphere

Denitrification can affect the environment unfavorably by increasing the N_2O content of the atmosphere. Enrichment of the air with N_2O and fluorocarbons threatens depletion of the ozone (O_3) layer in the upper atmosphere, thus allowing passage of more short-wavelength ultraviolet radiation and enhancing the incidence of skin cancer. Both NO and NO_2, produced as gaseous products of denitrification, either react in the soil or fail to persist in the lower atmosphere. However, N_2O is relatively stable in the troposphere and can diffuse into the stratosphere. There it photodissociates to form NO. The NO in turn catalyzes the destruction of O_3, which acts as a shield to incoming radiation. The significance of N_2O relative to other reactants destructive of O_3 is unknown.

The N_2O content of the atmosphere is now near 0.3 ppm. Currently

Table 9.5
Nitrogen Economy of Soil

Removals/losses	Range (%)
Crop uptake	0–60
Gaseous loss	0–30
Erosion	0–15
Immobilization	0–40
Leaching	0–10

about 65 megatons (metric) of fertilizer nitrogen are used annually. The total global nitrogen mineralized has been estimated to be 50 times as large as this. Without proper management, soil nitrogen that is mineralized is as susceptible to denitrification as added nitrogen. Possible counter-measures are (1) the use of nitrification inhibitors, (2) increased use of biological N_2 fixation, (3) optimization of both time and rate of fertilizer application to reduce losses, and (4) control of the mineralization and immobilization reactions to maintain as much as possible of the soil nitrogen in an available organic form. The denitrifying process in soil can serve both as a producer and as a consumer of N_2O. To the extent that atmospheric N_2O is carried back into strongly denitrifying soil by diffusion or precipitation, it becomes subject to conversion to N_2.

Uptake by Plant Roots

The ionic source of nitrogen preferred by organisms for protein synthesis is NH_4^+ rather than NO_3^-. Nitrogen in the reduced form can be incorporated readily into amino acids, whereas the oxidized form must first undergo reduction, the energy for which must be obtained from photosynthate. Preferential uptake of NH_4^+ over NO_3^- is easily demonstrable for microorganisms in laboratory cultures. In soils, the NH_4^+ ion is held on the exchange complex and its movement into and through the soil water is thus greatly restricted, whereas the NO_3^- ion can move to the plant root either by diffusion or by mass flow with water. Rapid nitrification of NH_4^+ can also make NO_3^- the actual ion source for the plant even though the anion is not added as such. Ease of movement to the root and ease of synthesis into amino acids and amides are not the only factors involved in plant uptake. Absorption of NH_4^+ depresses uptake of mineral cations, especially K^+, whereas NO_3^- depresses uptake of Cl^- and SO_4^{2-}. The residual or physiological acidity or alkalinity associated with anionic and cationic nitrogen sources can affect plant growth directly or through the effect of altered pH on the availability of other nutrients.

Even though the plant acquires only a part of the available mineral nitrogen, there are situations in which uptake outruns biosynthesis and NO_3^- accumulates within the plant. High NO_3^- accumulation can be expected when there is adequate NO_3^- in the soil and adverse growth conditions for the plants. Photosynthesis is required to supply energy for NO_3^- reduction and to provide organic acids for amino acid syntheses. Forages with high NO_3^- content are toxic to cattle and sheep; reduction of NO_3^- in the rumen leads to NO_2^- formation, which is toxic. Nitrite, whether formed in the rumen or prior to ingestion, becomes absorbed into the blood stream and there oxidizes oxyhemoglobin to methemoglobin, thereby blocking oxygen transport.

Nitrate Removal by Leaching

Leaching causes movement of NO_3^- from the upper soil profile to deeper soil layers and to ground waters. It is an abiotic process driven by both convection and diffusion. The equation commonly used for diffusion and convection acting jointly in the leaching of NO_3^- is as follows:

$$\frac{\partial c}{\partial t} = \overline{D} \frac{\partial^2 c}{\partial z^2} - v_0 \frac{\partial c}{\partial z}$$

where c is the concentration of NO_3^- (mg l^{-1}), $\overline{D}$ the apparent mean diffusion coefficient (cm^2 day^{-1}), v_0 the average pore velocity (cm day^{-1}; which is obtained by dividing the rate of flow of water by the volumetric water content of the soil), z the linear distance in the direction of flow (cm), and t the time (days).

The above equation is valid for a homogeneous soil and steady-state soil water conditions. It implies that flow rate is proportional to leaching volume and inversely proportional to the volumetric water content. In the complex fabric of soil, the parameters imposed are not constant spatially or temporally. The water movement velocity is not equal at all points; velocity near the center of a pore is greater than near the edge, and velocity through open channels (cracks, old root channels, animal burrows) is much greater than through a series of interconnected pores. Diffusion is reduced in water near particle surfaces; the layer of sorbed water at the surface is more viscous than water in the pore center.

In the field, the capability of a soil to release organic nitrogen for nitrification and the quantity and timing of fertilizer additions are largely responsible for excess NO_3^- in the cultivated layer. Past water regimes and management practices and current plant cover influence rate of water intake. Excess NO_3^- subjected to leaching moves downward in a wave or front. An example of NO_3^- movement into the deeper profile for a Canadian soil following its initial disturbance by cultivation is given in Fig. 9.3. This concentration of NO_3^- at depth only occurs in semiarid soils. Wetter soils often lose most of their NO_3^- to the groundwater during the noncropped or rainy season.

Nitrate that is leached to ground waters that are used as well waters poses a health hazard for infants, especially if the NO_3^- nitrogen content exceeds 10 ppm. Following consumption by an infant, the NO_3^- becomes

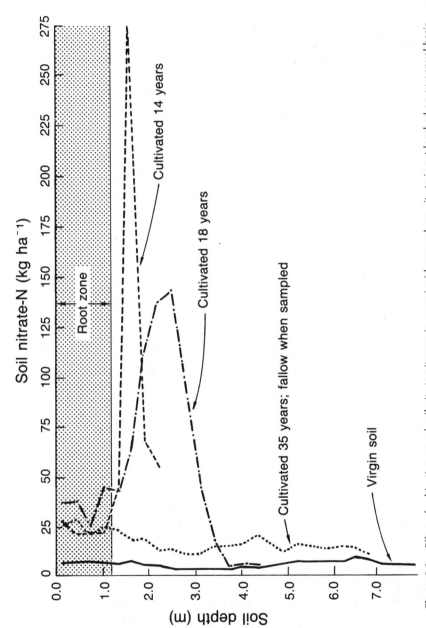

Figure 9.3. Effect of cultivation on subsoil nitrate-nitrogen in a semi arid zone where nitrate is not leached on an annual basis. From Campbell *et al.* 1975.

reduced to NO_2^-, which in turn irreversibly reduces the blood hemoglobin to methemoglobin. Death results from O_2 deficiency. The disease is known as methemoglobinaemia or blue-baby syndrome. Adults, having larger amounts of hemoglobin and usually lower water intake in proportion to body weight, are very seldom subject to the disease.

Erosional Losses of Nitrate

Nitrate is subject to removal off site by wind and water. During water erosion, NO_3^- removal does not parallel that of bulk soil removal. Because of its ready solubility, NO_3^- more easily enters into surface runoff than do particulates. Erosion decreases NO_3^- removal by leaching to the extent that it makes less NO_3^- available for leaching; by truncation of the soil profile and removal of topsoil, it lessens the amount of SOM available for mineralization and nitrification. Good conservation practices that lessen or eliminate loss of topsoil and NO_3^- may increase the amount of NO_3^- subject to leaching simply by preventing its loss by erosion. Good conservation increases infiltration and decreases runoff.

Nitrate may also be transported off site during wind erosion. Loss is limited to, and proportional to, particulate removal. The dust particles themselves, however, may not be truly representative of the topsoil. Drought usually precedes wind erosion, and during drought, capillary movement of water to the soil surface may enrich the soil surface with NO_3^-.

Localized Nitrate Accumulations

Nitrate that escapes both reduction (assimilatory and dissimilatory) and off-site transport can show massive accumulation within localized sites. Long-lasting accumulations occur in certain deserts, large caves, and small offshore islands, with bird dung (guano) serving as the source material. The most extensive guano deposit is located in the Atacama desert in Chile. This deposit was estimated to contain 250 megatons (metric) of $NaNO_3$. At one time Chilean exports of guano accounted for a major fraction of fertilizer nitrogen consumption. Accumulation of NO_3^- together with other salines also occurs in playas ("niter spots") in the western United States. These playa accumulations result from leaching of NO_3^- from rock parent material rather than from agricultural soils.

References

Aerssens, E., Tiedje, J. M., and Averill, B. A. (1986). Isotope labelling studies on the mechanism of N—N bond formation in denitrification. *J. Biol. Chem.* **261,** 9652–9658.

Campbell, C. A., Nicholaichuk, W., and Warder, F. (1975). Effect of a wheat summer fallow rotation on subsoil NO_3^-. *Can. J. Soil Sci.* **55,** 279–286.

Clark, F. E. (1962). Losses of nitrogen accompanying denitrification. *Proc. Int. Soil Conf. Palmerston North N.Z.,* pp. 173–176.

Cooper, G. S., and Smith, R. L. (1963). Sequence of products formed during denitrification in some diverse western soils. *Soil Sci. Soc. Am. Proc.* **27,** 659–662.

Firestone, M. K. (1982). Biological denitrification. *Agronomy* **22,** 289–326.

Knowles, R. (1981). Denitrification. *In* "Soil Biochemistry" (E. A. Paul and J. N. Ladd, eds.), Vol. 5. Dekker, New York.

Knowles, R. (1982). Denitrification in soils. *In* "Advances in Agricultural Microbiology" (N. S. Subba Rao, ed.), pp. 244–266, Butterworth, London.

Mosier, A. R., Hutchinson, G. L., Sabey, B. R., and Bayler, J. (1982). Nitrous oxide emissions from barley plots treated with NH_4, NO_3 or sewage sludge. *J. Environ. Qual.* **11,** 78–81.

Payne, J. W. (1981). "Denitrification." Wiley, New York.

Supplemental Reading

Focht, D. (1978). Methods for analysis of denitrification in soils. *In* "Nitrogen in the Environment" (D. R. Nielsen and J. G. MacDonald, eds.), Vol. 2, pp. 433–490. Academic Press, New York.

Jordan, C. F. (1985). "Nutrient Cycles in Tropical Forest Ecosystems." Wiley, New York.

Chapter 10

Return of Nitrogen to Soil: Biological Nitrogen Fixation

Global Distribution and Transfers of Nitrogen

Nitrogen is widely distributed on the earth in solid, dissolved, and gaseous forms. It accounts for only a very small percentage of the total weight of rocks and minerals, but because of the large mass of rocks in the lithosphere, the nitrogen contained therein represents 98% of the earth's total nitrogen. The largest potentially biologically available reservoir is the atmosphere; this is only equivalent to 1.2% of the earth's total nitrogen, but the gaseous N_2, accounting for 79% of the atmosphere, represents 3.9 $\times$ 10^{21} g (3.9 $\times$ 10^9 Tg) (Table 10.1).

Plants can accumulate nitrogen in their vegetation for periods of up to hundreds of years, as in trees, or cycle it seasonally in annual crops. An estimate of the amount of nitrogen absorbed annually by plants can be obtained from carbon photosynthetic values. Photosynthetic fixation of carbon on a global basis has been calculated to be 70,000 Tg. There is no single carbon:nitrogen (C:N) ratio for plant parts. However, an average value of 50:1 takes into account the relative distribution of trees, grasses, and cultivated crops. The nitrogen uptake value can therefore be reasonably accurately estimated at 1400 Tg. Plants are known to utilize only 30–60% of the available mineral nitrogen in soil. We can calculate a global soil mineralization capacity from the knowledge of nitrogen uptake and efficiency of uptake. At an efficiency of 40%, this nitrogen approximates 3500 Tg or 3% of the soil nitrogen content of 105,000 Tg. This value is higher than most previous estimates, but it takes into consideration two factors that usually are not considered: (1) the rapid remineralization of nitrogen in the tropics and (2) the fact that microbial uptake, or immobilization, of nitrogen is usually two to three times that of the mineral

Table 10.1

Global Nitrogen Contents and Fluxes for Soils

	Nitrogen	
	Content (Tg)[a]	Annual flux (Tg)
Atmospheric nitrogen	3.9×10^9	
Soil nitrogen	105,000	
Soil nitrogen mineralized		3500
Plant uptake		1400
Symbiotic N_2 fixation		120
Associative and free-living fixation		50
Fertilizer nitrogen		65
Fertilizer nitrogen utilized		26
Atmospheric inputs		25
Denitrification, volatilization		135
Runoff erosion		25
Leaching		93

[a] Each terragram (Tg) equals 10^{12} g or a million metric tons.

nitrogen accumulating during incubation of soil in the laboratory (net mineralization).

Symbiotic N_2 fixation at 120 Tg is low relative to soil nitrogen mineralization but represents 8.5% of the plant nitrogen uptake because of the direct transfer of the fixed nitrogen to the plant. Fertilizer nitrogen applications are estimated as equivalent to 50% of symbiotic fixation but because of lower uptake efficiencies represent only 2% of global plant nitrogen uptake. Losses of nitrogen shown in Table 10.1 include those from denitrification, volatilization, runoff, erosion, and leaching. The values for these processes vary greatly from site to site but on a long-term basis should approximately balance the nitrogen inputs, as they do in the table.

The air contains gaseous oxides, such as NO, NO_2, N_2O, HNO_2, and HNO_3. These range in concentrations from 0.5 to 2 ppb in regionally polluted areas; less than 0.1 ppb is found in areas remote from industrial pollution. The concentration of the gas NH_3 is more evenly distributed, with ranges of 2 to 3 ppb generally being found over most of the world. These gases can be absorbed by leaf surfaces and soil. In addition, the soil microflora has been shown to absorb and utilize gases such as N_2O. Relatively closed ecosystems, such as mature forests and grasslands, have low nitrogen losses and high internal recyling rates. These can therefore obtain a majority of their external needs from atmospheric inputs. The deposition rate of NH_3 and NO_3^- by rainfall depends on their concentration in the atmosphere and the amount of rainfall. Wet deposition of NO_3^-

nitrogen ranges from 5 kg ha^{-1} in polluted areas to less than 0.5 kg ha^{-1} in low-rainfall areas far from sources of industrial pollution. Ammonium nitrogen shows similar ranges; from 5 to 10 kg ha^{-1} year^{-1} are added to soils in industrial areas. As little as 0.5 kg ha^{-1} is added in remote areas, such as tundras. On a global basis these inputs are estimated at 25 Tg annually.

Disturbance of natural ecosystems leads to major losses of nitrogen, both to the atmosphere and to the ground water. Disturbed or cultivated areas, such as those employed in agriculture, require much larger inputs. The addition of fertilizer nitrogen, while not of major significance on a global basis, plays a very significant role in specific sites in intensive agriculture.

Organisms and Associations Involved in Nitrogen Fixation

The beneficial effects of leguminous plants are referred to in early Roman writings of more than 2000 years ago. Chinese authors during the same period also wrote about the beneficial effects of the use of the water fern, *Azolla*, in rice culture. A book on agricultural techniques written in AD 540, entitled "The Arts of Feeding the People" *(Chih Min Tao-Shu)*, further describes the cultivation and use of *Azolla* in rice fields.

Legumes used in rotation provided the agricultural stability that allowed the industrialization of western Europe. Sir Humphrey Davey, in 1813, was the first to suggest that nitrogen (azote) is derived from the atmosphere. He wrote, "peas and beans seem well adapted to prepare the ground for wheat." J. B. Boussingault, in his field rotation studies in France, published a series of papers from 1837 to 1842 that established the principles of nitrogen incorporation by legumes. Albert Thaer, in 1856 in Germany, wrote, "latterly the practice of sowing white clover with the last crop has become very general. Only a very few apathetic and indolent agriculturists or men who are firmly wedded to their opinions and customs neglect this practice."

The significance of nodules in symbiotic N_2 fixation was described by Helriegel and Wilfarth in 1886. Beijerinck, in 1888, isolated the organisms responsible for N_2 fixation by legumes and called them *Bacillus radicicola*. These were later to be renamed *Rhizobium*. The asymbiotic N_2-fixing organism, *Clostridium*, was isolated by Winogradsky in 1890, and the aerobic organism, *Azotobacter*, was described by Beijerinck in 1901. Thus by the end of the century, nearly all the major N_2-fixing organisms had been characterized.

Diverse group of prokaryotes contain the enzyme nitrogenase respon-

Table 10.2
Organisms and Associations Involved in Dinitrogen Fixation[a]

Agrobacterium Association type		Representative prokaryotic genera
Organotrophs	Aerobic	_Azotobacter, Beijerinckia, Derxia, Xanthobacter, Rhizobium_
	Facultative Aerobic	_Bacillus, Klebsiella, Azospirillum, Thiobacillus_
	Anaerobic	_Clostridium, Desulfovibrio, Desulfotomaculum_
	Genetically engineered	_Salmonella, Escherichia, Serratia_
Free-living phototrophs	Cyanobacteria	_Nostoc, Trichodesmium, Anabaena, Gloeothece_
	Purple nonsulfur bacteria	_Rhodopseudomonas, Rhodospirillum_
	Purple and green sulfur bacteria	_Chromatium, Chlorobium, Thiocapsa_
Organotrophs	Rhizosphere	_Azospirillum, Azotobacter, Bacillus_
Nonnodule nodule	Phyllosphere	_Klebsiella, Beijerinckia_
	Legume	_Rhizobium_
	Nonlegume	_Rhizobium_
	Nonlegume, Actinomycete	_Frankia_
	Nonlegume, _Gunnera_	_Nostoc_
Associative phototrophs	Lichens	_Nostoc, Stignonema, Calothrix_
	Liverworts	_Nostoc_
	Mosses	_Halosiphon_
	Gymnosperms (_Cycas_)	_Nostoc_
	Water ferns (_Azolla_)	_Anabaena_
	Endocynoses (_Oocystis_)	_Nostoc_

[a]Adapted from Havelka _et al._ (1982).

sible for the fixation of N_2. These bacteria are now called diazotrophs and include organotropic bacteria, phototrophic sulfur bacteria, and cyanobacteria (blue-green algae) (Table 10.2).

Free-Living Organotrophs

The aerobic, free-living, N_2-fixing bacteria that utilize organic substrates as a source of energy include *Azotobacter,* found in neutral and alkaline soils. Somewhat similar bacteria, *Beijerinckia* and *Derxia,* have a broader pH range and are more often found in acidic soils, especially in the tropics. Genetic analysis shows *Beijerinckia* to be more closely related to another N_2-fixing bacterium, *Azospirillum,* than to *Azotobacter.* These organisms, as well as the N_2-fixing *Klebsiella,* are found growing in humid environments on leaf surfaces or in leaf sheaths (the phyllosphere), as well as in the soil and on root surfaces.

Azotobacter, Beijerinckia, and *Rhizobium* require aerobic conditions for the production of the extensive energy required for N_2 fixation. However, in these organisms as in all other diazotrophs, the activity of the enzyme nitrogenase is inhibited by O_2. Special mechanisms for the protection of nitrogenase include the association of the N_2-fixing complex with membranes within the cell. Slime production and clump formation by the cells also provide protective mechanisms. This reduces O_2 diffusion rates to the enzyme. Another feature of aerobic N_2-fixing bacteria is the high level of respiration within the cells. This in *Azotobacter* helps protect the enzyme from O_2 by maintaining low O_2 concentrations. It also has been suggested that in the presence of O_2 the nitrogenase enzyme can undergo a conformational change leading to a protected but non-N_2-fixing enzyme. This change is reversible, and the enzyme can return to its N_2-fixing state when intracellular O_2 no longer is present.

Facultative microaerophilic organisms such as *Klebsiella, Azospirillum,* and *Bacillus* produce energy in the form of ATP by oxidative pathways in an environment where nitrogenase does not need to be as well protected from O_2, as is the case with *Azotobacter.* Anaerobic diazotrophs such as *Clostridium* and the sulfate reducers, including *Desulfovibrio* and *Desulfotomaculum,* also use organic compounds as electron donors. The fermentative pathways of these organisms, however, lead to the buildup of organic intermediates and result in only low amounts of energy being available for N_2 fixation. However, there are certain environmental conditions in which high substrate availability combined with anaerobic conditions, such as waterlogging, have been shown to result in extensive N_2 fixation by these organisms. The anaerobic conditions lower the potential for energy oxidation from substrates. They, however, also lower the competitive use of this substrate by other organisms, such as aerobic bacteria and fungi.

The amount of N_2 fixed by the free-living diazotrophs and by the phyllosphere organisms is generally only a few kilograms per hectare. While not large on an individual–area basis, without exception nearly all habitats investigated have shown some fixation. The impact of the free-living diazotrophs is significant when considered on a global basis and in areas where internal cycling of nitrogen together with slow growth rates leads to low requirements for external inputs.

Free-Living Phototrophs

Cyanobacteria (blue-green algae) were shown by Drewes in 1928 (cited in Fogg, 1977) to fix N_2. They occur in many aquatic situations. They also are found at or just below the surface in many soils, especially those that have been newly exposed by erosion or landslides. Cyanobacteria are common in many aquatic areas, such as rice paddies and in small lakes, with moderate nutrient levels. Here they can at times develop blooms that produce animal neurotoxins of high potency. They are the only prokaryotes that exhibit photosynthetic mechanisms characteristic of higher plants, in that they contain photosystem II and evolve O_2, yet they can survive in low-light, low-O_2 environments.

The free-living diazotrophic cyanobacteria have three morphological forms. The unicellular cyanobacter *Gloeothece* (formerly known as *Gloeocapsa*) is an aerobic soil organism sensitive to high light. It protects its nitrogenase by mechanisms similar to those described for *Azotobacter* in that the enzyme is protected by membranes that separate it from the O_2 produced during photosynthesis.

The ability to fix N_2 is especially prevalent in filamentous cyanobacteria containing specialized cells called heterocysts (Fig. 10.1). The thick-walled heterocysts, which occur every 10 to 15 cells in the filaments, protect the nitrogenase from O_2 damage by physical means, such as membranes, to prevent gas diffusion. The absence in the heterocysts of photosystem II, the O_2-evolving photosynthetic subsystem, provides further protection. *Nostoc*, which is an example of the heterocystous filamentous type, appears as a crust on many grassland and desert soils. The crust may be dry during most of the time but it can grow and fix N_2 while moistened by a short rainfall or morning dew. The crusts have a high requirement for calcium and thus prefer calcareous soils; they also often respond to inorganic phosphorus additions, either as fertilizer phosphorus or the phosphorus produced during a fire. The third morphological type includes the nonheterocystous filamentous *Plectonema* (Table 10.3).

Figure 10.1.   (a-e) Electron micrographs of vegetative cells (V) and heterocysts (H) of *Anabaena cylindrica*. Heterocysts have a thick cell envelope (E), and pores (P) between cells and heterocyst. Note dense osmiophilic plug (PP) and contorted lamellae (CL). Polyphosphate bodies (PB) are found in vegetative cells. (From Stewart, 1977.)

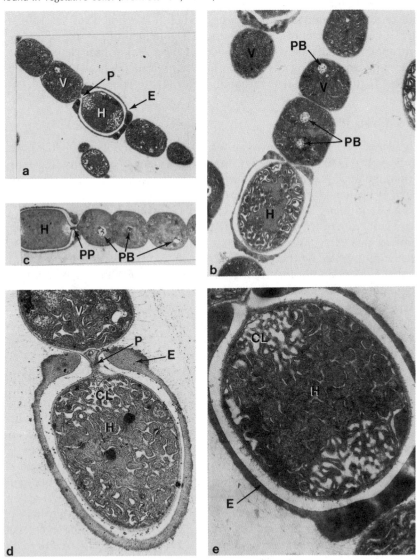

Table 10.3
Dinitrogen Fixation among Cyanobacteria

Order	Family	Genus and number of species (in parentheses) implicated in N_2 fixation
Chroococcales (circa 35 genera)	Chroococcaceae	*Gloeothece* (1)
Oscillatoriales (circa 100 genera)	Nostocaceae	*Anabaena* (12)
		Anabaenopsis (1)
		Aphanizomenon (1)
		Aulorisa (1)
		Chlorogloea (1)
		Cylindrospermum (4)
		Gloeotrichia (1)
		Nodularia (1)
		Nostoc (8)
		Trichodesmium (1)
	Rivulariaceae	*Calothrix* (5)
	Scytonemataceae	*Scytonema* (2)
		Tolypothrix (1)
	Oscillatoriaceae	*Oscillatoria* (1)
		Plectonema (1)
		Schizothrix (1)
Stigonematales	Stigonemataceae	*Fischerella* (2)
		Hapolosiphon (1)
		Mastigocladus (1)
		Stigonema (1)
		Westelliopsis (1)

Rice paddies, the best-studied site of N_2 fixation by cyanobacteria, contain *Anabaena, Gleotrichia,* and *Scytonema*. These organisms have been credited with the long-term stability of rice culture over thousands of years. The measurements of N_2 fixation under such conditions, however, include contributions from both the free-living organotrophs and the free-living phototrophs. As in all N_2 fixation measurements, the amounts fixed depend on environmental conditions, such as moisture, available nutrients, and the period over which N_2 fixation is effective. Estimates of N_2 fixation range from the average of 30 kg ha^{-1} $year^{-1}$ shown in Table 10.4 to as high as 70 kg ha^{-1} $year^{-1}$.

The phototrophic N_2-fixing bacteria other than cyanobacteria are aquatic, marine, and freshwater organisms in which H_2O does not function as the electron acceptor during photosynthesis; thus O_2 is not evolved. Alternate electron acceptors, such as H_2S and sulfur, are utilized. Although of diverse morphology (Table 10.2), these organisms are placed in the order Pseudomonadales. They are often found at the sediment–water interface of lakes, where light can penetrate to an environment that is held at a low redox potential in the presence of reduced sulfur compounds. The green and purple sulfur bacteria are photolithotrophs using H_2S as an electron donor and reducing CO_2 in the presence of light. The purple nonsulfur bacteria *(Rhodospirillum)* require organic compounds as the electron donor during photosynthesis and thus are photoorganotrophs. These organisms also can use organic compounds as a source of energy. Although their photosynthesis must occur under anaerobic conditions, the bacteria can grow under aerobic conditions as chemorganotrophs. Their contribution to N_2 fixation can be significant under conditions in which extensive blooms occur.

Diazotrophs Associated with Grasses

Plant surfaces provide a source of carbon, moisture, and a partially protected habitat for the development of otherwise free-living diazotrophs. The leaf surface, or phyllosphere, contains aerobic and microaerophilic organotrophs as well as cyanobacteria. The most commonly isolated types in temperate environments are *Klebsiella* species; *Azotobacter* and *Beijerinckia* are common on tropical plants, especially those in moist environments.

The root zone, or rhizosphere, was shown many years ago to be an active site for N_2 inputs. Dobereiner highlighted these associations when she found extensive growth and high N_2 inputs with the C_4 Brazilian grass *Paspalum notatum* growing in association with *Azotobacter paspali* (Dobereiner and Day, 1975). Later work rediscovered the previously described diazotrophic spirilli now named *Azospirillum*. Two species of *Azospirillum* are associated with a diverse range of plant hosts, such as the salt-marsh grass *Spartina* as well as sugar cane, rice sorghum, maize, millet, and wheat.

The diazotrophs grow in the rhizosphere immediately adjacent to the roots (Fig. 10.2) and invade the root cortex. Nitrogen fixation rates vary from negligible levels to 20 kg ha^{-1} year^{-1} in upland grasses, perhaps even higher in rice. These diazotrophs also produce growth hormones that may affect root development. Their contribution to plant growth may be due partly to mechanisms other than the fixed nitrogen they add to the association. A portion of the variability in their N_2 fixation under aerobic conditions may be explained by the observation that vesicular–arbuscular mycorrhizal fungi, such as *Glomus,* compete for plant carbon with the diazotrophic *Azospirillum*. Nitrogen fixation has been shown to be lowered

Figure 10.2. Scanning electron micrographs of *Azobacter brasilense* absorbed to root hairs of maize (top, from Patriquin, 1982) and millet (bottom, photo courtesy of F. Dazzo).

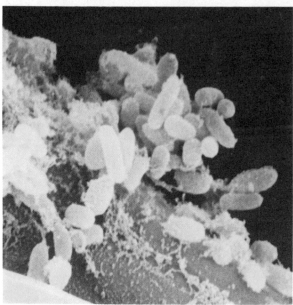

in plants heavily infested with *Glomus*. High soil phosphorus levels, which inhibit mycorrhizas, may thus enhance grass-associated fixation.

Legumes

Legumes represent a group of plants comprising 700 different genera with 14,000 species. Nodule presence and N_2 fixation have not been determined in many of these, especially those in tropical forests. The 100 agriculturally important legumes growing on some 250×10^6 ha of land contribute a significant portion of the biological N_2 fixation (Table 10.4). The estimate of 140 kg ha^{-1} year^{-1} shown in Table 10.4 has been criticized as being too high for an average value. Although certain fields, as in New Zealand, can fix 800 kg ha^{-1} year^{-1}, many legumes grow on arid soils and show negligible fixation rates. In addition, a large portion of the legume crop of Europe now receives fixed nitrogen in the form of fertilizers. Grain legumes, such as peas, beans, and soybeans, also are known to have much lower fixation rates than the forage legumes, such as alfalfa. The grain legumes often depend on soil nitrogen for at least 50% of their nitrogen requirements.

Table 10.4
Dinitrogen Fixation in Various Land-use Types[a]

Land use	ha ($\times 10^6$)	Nitrogen Fixed (kg ha^{-1} year^{-1})	Tg
Legumes	250	140	35
Rice	135	30	4
Other cultivated crops	1,015	5	5
Permanent meadows, grasslands	3,000	15	45
Forest and woodland	4,100	10	40
Unused	4,900	2	10
Ice covered	1,500	0	0
Total land	14,900		139
Sea	36,100	1	36
	51,000		175

[a]Adapted from Burns and Hardy (1975).

The diazotrophs that form associations with the major agriculturally important legumes are shown in Table 10.5. The slow-growing *Bradyrhizobium lupini* and *B. japonicum* have been placed in the same classification as the cowpea group, a promiscuous group of rhizobia that cross-inoculate with a broad range of hosts. Acid production from carbohydrates, in addition to the other characteristics, such as growth rate, separate the fast from the slow-growing rhizobia. The taxonomy of symbiotic bacteria associated with legumes is now undergoing change. Diagnostic features such as serology, DNA, homology, and rescue of genetic markers on chromosomal genes are now used for classification of the strains shown in Table 10.5. These do not include many types of *Rhizobium,* such as those found on trees in the tropics, nor are they the best representation from a genetic basis. The classification, however, is useful for purposes such as the production of inocula.

A nonleguminous shrub, *Parasponia* (previously known as *Trema*), in the family Ulmaceae has been shown to be nodulated with effective *Bradyrhizobium.* This suggests the possibility of associations between this important, active diazotroph and other plant hosts. Other interesting associations have been found. Soil nitrate nitrogen inhibits the formation of nodules and of nitrogenase. Pea and soybean cultivars without nitrate reductase have been isolated. These are capable of fixation in the presence of nitrate. However, such hosts have usually been found to have poor growth capacity. Another leguminous plant, *Sesbania rostrata,* has the capability of forming nodules on both the roots and the stem. This allows

Table 10.5
Classification of the Symbiotic Bacteria of Agriculturally Important Legumes[a]

Major classification	Species	Plant hosts
Fast growing		
Rhizobium	*meliloti*	*Medicago* (alfalfa), *Melilotus, Trigonella*
	trifolii	*Trifolium* (clover)
	leguminosarum	*Pisum* (pea), *Vicia, Lathyrus*
	phaseoli	*Phaseolus* (bean)
Slow growing		
Bradyrhizobium	*lupini*	*Lupinus, Lotus*
	japonicum	*Glycine* (soybean), some cowpeas
	spp. "cowpea"	*Vigna* (cowpea), others
	parasponiae	*Parasponia,* cowpea, others

[a]Adapted from "Bergey's Manual of Determinative Bacteriology" 9th ed. (1983).

the plant to fix N_2 when growing under wet conditions, where root nodules are not developed. In addition, the nodules on the stem appear to have the capacity for fixing N_2 even though roots are absorbing nitrate from the soil.

Biotechnological techniques will, it is hoped, produce new plant–microbe combinations that have high N_2 fixation capabilities in the presence of soil mineral nitrogen under a wider range of environmental conditions. Grain legumes, such as peas, dry beans, and soybeans, which now do not have adequate N_2-fixing potential, should also be improved.

Actinorhizal Associations

The actinomycete *Frankia* forms nodules in association with 170 plant species belonging to 17 genera, 8 families, and 7 orders (Table 10.6). These perennial woody shrubs and trees, of worldwide distribution, are as important to forest and wild lands as legumes are to agriculture. The best-known association for *Frankia* is the *Alnus*-type nodule in alders. This is characterized by a coralloid root nodule with dichotomous branching (Fig. 10.3). These nodules are pink. However, this color is not produced by the oxygen-controlling protein, leghemoglobin, as in *Rhizobium* nodules, but is attributed to anthocyanin pigments. The other major nodule structures developed by *Frankia* are those found in nodules of the *Myrica–Casuarina* type. The apex of each nodule produces an upward-growing root such that the nodule mass is covered with upward-growing roots (Fig. 10.3). This upward growth pattern facilitates gas movement from the atmosphere to the N_2-fixing endophyte under the low O_2 tensions found in the high soil moisture conditions that these plants often encounter.

The associations involving actinomycete N_2 fixation occur in habitats ranging from arctic to tropical, and semidesert to rainforest. This is reflected in the estimates of N_2 fixation ranging from 2 to 300 kg ha^{-1} year^{-1} (Table 10.6). *Casuarina*, a woody species, has special adaptations to tropical conditions and is an important fuel-wood plant in many tropical areas. *Myrica*, usually found in boglike environments, is also common to landslide areas, eroded slopes, and mining wastes on the eastern coast of North America.

Alders range in size from shrubs to large trees. They tend to grow on stream banks or in high-rainfall environments, such as the Pacific Northwest of the United States, and also have been shown to fix up to 300 kg nitrogen ha^{-1} year^{-1}. Soils associated with alders usually have high organic matter contents and at certain times of the year also develop high nitrate

Table 10.6
Classification of Nonleguminous Dinitrogen-Fixing Angiosperms Producing Symbioses with *Frankia*[a]

Order	Family	Genus	Number of nodulated species (in parentheses, total number of species)	Estimates of field N$_2$ fixation (kg ha^{-1} year^{-1})
Casuarinales	Casuarinaceae	*Casuarina*	24 (45)	230
Myricales	Myricaceae	*Myrica*	26 (35)	3–25
		Comptonia	1 (1)	26–30
Fagales	Betulaceae	*Alnus*	33 (35)	26–300
	Eleagnaceae	*Eleagnus*	16 (45)	15
		Hippophae	1 (3)	2–180
		Shepherdia	3 (3)	60
Rhamnales	Rhamnaceae	*Ceanothus*	31 (55)	—
		Discaria	6 (10)	—
		Colletia	3 (17)	—
		Trevoa	1 (6)	—
Coriariales	Coriariaceae	*Coriaria*	13 (15)	—
Rosales	Rosaceae	*Rubus*	1 (250)	—
		Dryas	3 (4)	18–60
		Purshia	2 (2)	—
		Cercocarpus	4 (20)	—
Cucurbitales	Datiscaceae	*Datisca*	2 (2)	—

[a] Adapted from Becking (1982).

levels. Nitrogen fixation in alders, as is the case with highly productive fields of legumes, results in the production of soil acidity. This acidity is only an indirect effect of the N$_2$-fixing process and the higher subsequent N mineralization. Soils supporting alder therefore are a good area to study the effects of acidity, such as that produced by acid rain. Agricultural soils supporting N$_2$ fixation, if not highly buffered, will require pH adjustment by liming.

The shrubs *Shepherdia* and *Eleagnus* (wolf willow) are well known on the Great Plains of North America. They also are extensively found in areas of active revegetation. Such areas can be found on the new soils created by the retreat of glaciers in Alaska and in the revegetation of mine spoils. The present grasslands of North America do not have an extensive legume component. It is possible that these actinorhizal shrubs played a

Figure 10.3.   *Alnus* and *Myrica/Casuarina* types of root nodules. (A) Coralloid root nodule of *Alnus glutinosa*. (× 0.5.) (B) Detached and divided root nodules of *A. glutinosa,* showing dichotomous branching of nodule lobes. (× 0.9.) (C) *Casuarina equisetifolia* root nodules, showing the feature that the apex of each nodule lobe gives rise to a negatively geotrophic root. (× 0.9) (From Becking, 1982).

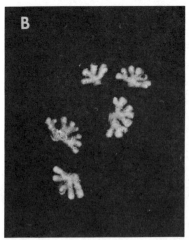

major role, in the past, in the accumulation of the large amounts of nitrogen now present in these soils.

Three genera of Rosaceae, of the 100 genera within this family, are capable of forming nodules. These are *Dryas, Purshia,* and *Cercocarpus. Dryas* is a low-growing nodulated shrub of Alaska and Eurasia, whereas *Purshia tridenta* is an important browse plant in arid regions of the western United States. *Ceanothus* is the predominant N_2-fixing shrub of the fire-type scrub associations known as chaparral in the Mediterranean-type grassland and mixed forest areas of the American West. It can add significant amounts of fixed nitrogen but can also compete with replanted trees for available water.

The cyanobacter *Nostoc* appears to act as an organotroph when associated with the herbaceous dicot, *Gunnera.* This involves morphological adaptation of the invaded stem tissue by *Nostoc* to produce a structure known as a nodule or gland. The genus *Gunnera* includes plants ranging from herbs to 1.5-m-tall, stout shrubs. These live in wet environments in South America, New Zealand, Southeast Asia, and Africa. The cyanobacteria within the glands have a higher frequency of heterocysts than does free-living *Nostoc.* Fixation of N_2 in the absence of $^{14}CO_2$ incorporation into the *Nostoc* gland has led to the interpretation that *Nostoc* is organotrophic in this association. Estimates of N_2 fixation range from 10 to 72 kg ha^{-1} year^{-1}.

Phototrophic Associations

The association of algae with fungi results in the formation of lichens. The symbiotic associations are recognized as taxonomic identities. Lichens that fix N_2 have cyanobacteria as the photosynthesizing symbiont (the phycobiont). The cyanobacter appears to obtain physical protection and

Table 10.7
Representative Dinitrogen-Fixing Lichens[a]

Lichen	Cyanobacterium	Lichen	Cyanbacterium
Collema	*Nostoc*	*Nephroma*	*Nostoc*
Dendriscocaulon	*Scytonema*	*Peltigera*	*Nostoc*
Epheba	*Stigonema*	*Solorina*	*Nostoc*
Lepogium	*Nostoc*	*Stereocaulon*	*Nostoc*
Lichina	*Calothrix*	*Sticta*	*Nostoc*
Lobaria	*Nostoc*	*Placynthium*	*Dichothrix*

[a]Adapted from Millbank (1977).

nutrients, whereas the fungus (the mycobiont) obtains carbon and, often, nitrogen from the phycobiont. Those lichens shown to be N_2 fixers are shown in (Table 10.7). Annual growth increments of many lichens can be as little as 1 mm. More rapid growers, such as *Peltigera*, show an increment of 2 to 3 cm $year^{-1}$. *Peltigera* is widely distributed and of special significance in the Arctic, where other nitrogen inputs are low. The effect of acid-forming lichens on the initial weathering of rocks and the establishment of vegetation on newly exposed surfaces also is worth noting.

Figure 10.4.   *Azolla.* Longitudinal section through a young, functional macrosporacarp of *A. filiculoides*. The algal cells are still in filaments. Centrally in the cavity a cross-section through the developing macrosporangium is visible. The symbiotic algae are equally distributed over the total free space of the macrosporocarp. (× 360.) (From Becking, 1978.)

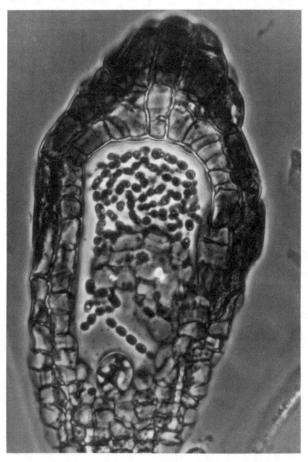

The freshwater fern, *Azolla*, floating on the surface of quiet waters, such as drainage ditches and rice paddies, is of worldwide distribution. It is most abundant in the tropics, where, if undisturbed by winds, it can grow to a heavy carpet. The four identified plant host species contain a heterocystous cyanobacterium *(Anabaena)* that can supply the *Azolla* with a large percentage of its total nitrogen requirement.

The *Anabaena* filaments occupy a cavity in the ventral surface of the multibranched floating stem (Fig. 10.4). Nitrogen fixation requires a high content of soluble iron and phosphorus in the water. Only rarely do the *Azolla* roots reach the sediment. The *Azolla–Anabaena* complex also is heavily grazed by a number of invertebrates, and optimum N_2 fixation requires control of these predators. Nitrogen inputs of 100 kg ha^{-1} have been found in a single rice crop, and 330 kg ha^{-1} has been found over a 222-day growing period. *Anabaena* releases 50% of the nitrogen fixed as soluble NH_4^+. *Azolla* can be grown either in conjunction with rice, as a cover, or as a green manure crop during periods when the fields are not

Figure 10.5. Yields of rice grain for unfertilized control, *Azolla* cover (dual culture with rice), *Azolla* green manure, and combined green manure and cover treatments. *Azolla* green manure represented 60 kg nitrogen ha^{-1} of decomposing *A. filiculoides* cover grown in dual culture, with rice representing 31 kg nitrogen ha^{-1}. *Azolla mexicana* cover represented 38 kg nitrogen ha^{-1}. (After Talley *et al.;* from Peters and Calvert, 1982.)

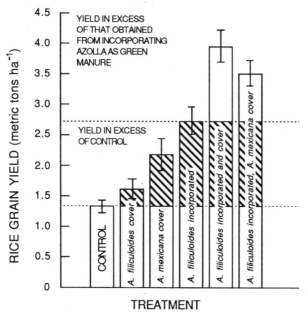

in rice. Approximately 40% of the rice fields in northern Vietnam utilize one of these two systems (Fig. 10.5). The yield of rice without *Azolla* (control) was 1.3 metric tons. This was doubled by the incorporation of *Azolla* as green manure. Another 1.3 metric tons were obtained when fields that had previously been green manured were reinoculated with *Azolla* as a cover treatment in conjunction with the rice crop. The major constraint of *Azolla* involves the labor required. It must be propagated, transported, and maintained in storage in the vegetative state. More research is required on mechanisms of inoculation, on maintenance of the iron and phosphorus supply in the water, on pest control, and on mechanisms to initiate spore formation.

Nodulation and Nitrogen Fixation in Legumes

Establishment of the legume–*Rhizobium* symbiosis involves interaction of both symbionts. Rhizobia are attracted to root surfaces, where they proliferate and attach to root cells by specific mechanisms that involve the interaction of complementary macromolecules. These macromolecules consist of carbohydrate (glyco-) proteins called lectins, present on legume roots. Compatible rhizobial symbionts produce an extracellular, acidic polysaccharide that interacts selectively with the plant lectin. There is a bidirectional transfer of information from the bacterium to the plant and from the plant to the bacterium (Fig. 10.6). This explains some of the host–bacterium specificity associated with nodulation.

The initial response of the root hair to *Rhizobium* often involves a tight curling of the root-hair tip into a form resembling a shepherd's crook. An infection thread is formed by enzymatic dissolution of the root-hair cell wall. The plant then synthesizes new material to encase the invading rhizobia. The plant cell nucleus controls the ingrowth of the infection thread. The base of the hair cell is usually reached 48 hr after infection.

Bacteria of the promiscuous cowpea group, belonging to the genus *Bradyrhizobium,* nodulate a broad range of legumes (many of which are tropical) without the specificity encountered in the common legumes of temperate areas. These bacteria invade through openings formed at the site of root-hair emergence.

The release of bacteria into cortical cells, either from the infection thread or through direct penetration of intercellular spaces, is followed by rapid proliferation of the bacteria within the host cells. The bacteria within the cells lose their rod-shaped structures and become pleomorphic, club-shaped cells (Fig. 10.7). They also develop the enzyme complexes required for N_2 fixation. These forms of the bacteria are called bacteroids. Up to

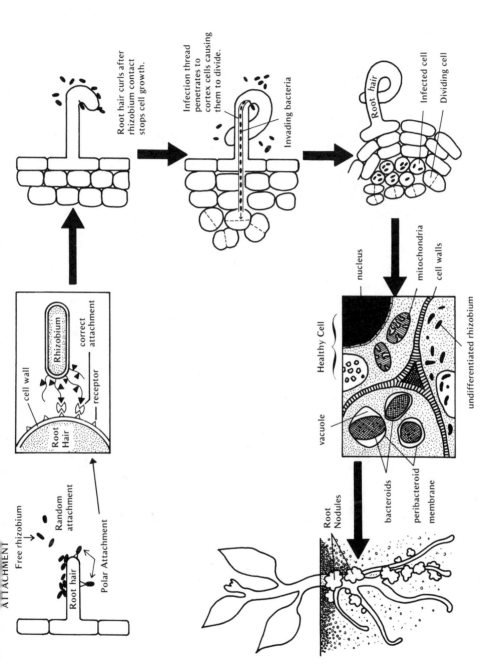

Figure 10.6. Stages in the infection of legume roots by rhizobia. (From Ahmadjian and Paracer, 1986).

Figure 10.7.   Scanning electron micrograph of *Rhizobium* on soybean root, showing bacteroidal cells; inset shows five cells in a membrane envelope. Scale bar, 2 μm. (From Tu, 1975.)

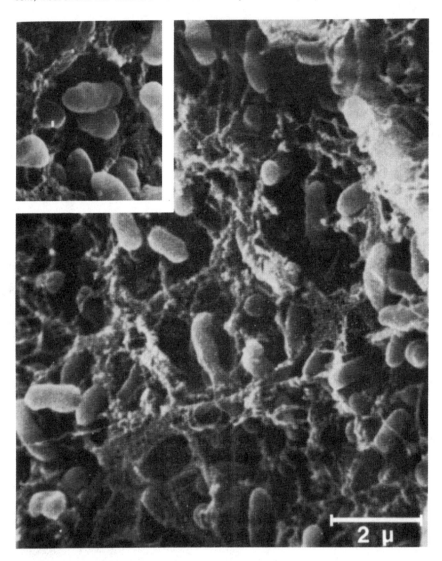

10,000 bacteroids can be found per root cell in an active nodule, composed of approximately equal parts of bacterial and host plant tissue.

The nodule is compartmentalized by plant membrane envelopes. Although the bacteroid requires O_2 for the production of the high levels of ATP required in N_2 fixation, the N_2-fixing enzyme, nitrogenase, is extremely O_2 labile. The level of O_2 in nodule tissue is controlled by an iron–heme protein that is pink, similar to the hemoglobin of blood, and is thus called leghemoglobin. Nodules of the nonlegume *Parasponia* and those of the actinorhizal shrubs do not contain leghemoglobin. The mechanism of O_2 control in those associations is not clearly understood.

Biochemistry of Nitrogen Fixation

The mechanism of N_2 fixation appears to be quite similar in most of the N_2-fixing prokaroytes. The nitrogenase complex consists of two iron–sulfur proteins. One of these is designated the iron protein, with a molecular weight of 60,000. This contains 4 iron and 4 labile sulfur atoms in a cubic structure. Spectroscopic evidence indicates that the 4-iron/4-sulfur cluster operates between oxidation states 2 and 3 within the iron protein. The second and larger iron–sulfur protein contains between 28 to 32 iron and 28 labile sulfur atoms as well as 2 molybdenum atoms per molecule. The molecular weight of this molybdenum–iron–sulfur protein varies between 200,000 to 300,000. For nitrogenase to function, both components of the complex must be present. These two proteins catalyze the reduction of N_2 to NH_3. They also reduce C_2H_2 to C_2H_4, and H^+ to H_2, as well as the other reactions shown in Table 10.8.

Table 10.8
Nitrogenase-Mediated Reactions[a]

Name	Formula	Products
Dinitrogen	$N\equiv N$	$NH_3 + H_2$
Acetylene	$HC\equiv CH$	$H_2C=CH_2$
Cyanide	$[C\equiv N]^-$	$CH_4 + NH_3$
Allene	$H_2C=C=CH_2$	$H_3C-CH=CH_2$
Cyanogen	$N\equiv C-C\equiv N$	CH_4
Azide	$\overset{+}{N}\equiv N-\overset{-}{N}$	$N_2 + NH_3 + N_2H_4$
Nitrous oxide	$\overset{+}{N}\equiv N-\overset{-}{O}$	$N_2 + NH_3$
Hydrogen ion	$[H]^+$	H_2

[a]Adapted from Postgate (1982).

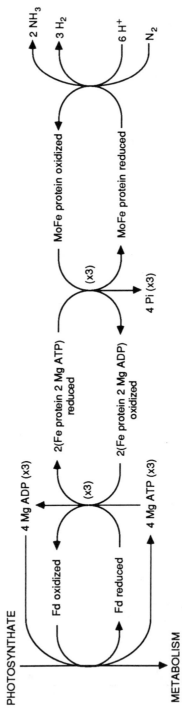

Figure 10.8. Overall reaction in dinitrogen fixation. (Fd, ferredoxin; Mg, magnesium. (From Atkins and Rainbird, 1982.)

Electrons flow through normal metabolic routes to ferrodoxin and then to the iron protein (Fig. 10.8). This, when complexed with ATP, reduces the iron–sulfur protein, with the formation of ADP and the release of inorganic phosphate (P_i). The protein, iron–sulfur called nitrogenase, passes the electrons to N_2 or one of the reducible substrates shown in Table 10.8. Activity of the molybdenum–iron enzyme during N_2 fixation also results in the coreduction of H^+ to H_2. The proportion of H^+ to the amount of N_2 reduced varies. At low electron flux, H_2 is the predominant product; at high electron flux and high ATP content, N_2 is an effective competitor and a high percentage of NH_3 is formed.

Variability in the amount of ATP hydrolyzed per electron pair transferred and in the ratio of N_2 to H^+ reduced makes it difficult to establish an overall reaction equation. Under optimum laboratory conditions with a purified enzyme complex, a value of 2 ATP per electron passing to reducible substrate has been demonstrated. The N_2 reduction process has been shown by Atkins and Rainbird (1982) to be

$$N_2 + 6\,e^- + 12\,ATP + 8\,H^+ \xrightarrow{Mg^{2+}} 2\,NH_4^+ + 12\,ADP + 12\,P_i$$

The associated reduction of H^+ in enzyme complexes indicates equal reduction of N_2 and H^+, as follows:

$$2\,H^+ + 2\,e^- + 4\,ATP \xrightarrow{Mg^{2+}} H_2 + 4\,ADP + 4\,P_i$$

The overall equation then becomes

$$N_2 + 16\,ATP + 8\,e^- + 10\,H^+ \xrightarrow{Mg^2} 2\,NH_4^+ + H_2 + 16\,ADP + P_i$$

Hydrogen evolution is a constant feature of nitrogenase activity. In some organisms the presence of the enzyme hydrogenase results in recapture of some of this lost energy and the production of ATP for use in cell reactions. Figure 10.9 shows the overall reaction in the nodule. This figure relates the input of photosynthate through electron transport and oxidative phosphorylation to the nitrogenase. It also shows that the action of hydrogenase does not alter the N_2-fixing reaction but can result in recapture of some of the energy that otherwise would be lost from the complex as H_2. Studies with plants nodulated with *Rhizobium* having H_2-uptake activity (Hup^+) relative to those that do not (Hup^-) indicate that the presence of hydrogenase in specific rhizobia results in a significant increase in plant nitrogen content.

The first product of fixation is NH_4^+. Ammonium has a repressive effect on the production of new amounts of nitrogenase in both free-living and symbiotic N_2 fixers. High levels of glutamine formed from NH_4^+ cause

Figure 10.9. Diagram showing the postulated relationships between hydrogen metabolism and nitrogen fixation within a legume module. (After Schubert and Evans, 1976.)

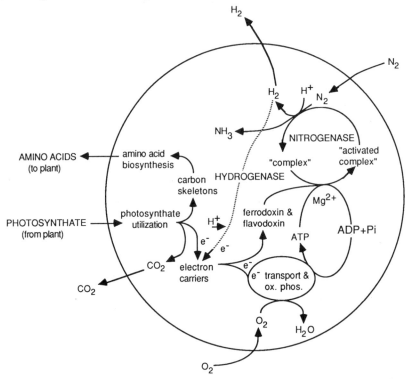

an adenylation of the NH_4^+-assimilating enzyme, glutamine synthetase (Fig. 10.10). This in turn affects the coding for the production of new nitrogenase in the bacteroid.

Nitrate, when present in the soil, has an inhibitory effect on both nodulation and N_2 fixation. The mechanisms of inhibition are not clearly understood. Nitrate assimilation requires large amounts of organic substrate and nitrate reduction could deprive the N_2-fixing system of an available carbon supply. The alternate theory is that the first reduction product of NO_3^- which is NO_2^-, could complex with leghemoglobin and destroy its oxygen-regulating capabilities. Legume plants without nitrate reductase activity have been obtained for experimental purposes. Theoretically, these plants should be able to fix N_2 in the presence of soil NO_3^-.

Oxygen diffusion rates into the nodules of legumes affect the extent of N_2 fixation. Waterlogging, even for short periods of time, is detrimental to nodule functioning. Atmospheric O_2 concentrations greater than the

Figure 10.10. Pathways of ammonia assimilation in temperate legumes. GDH, Glutamate do-hydrogenase; GOGAT, glutamine–emide 2-oxaloglutarate amino transferase; GS, glutamine synthetatase. (From Child, 1981.)

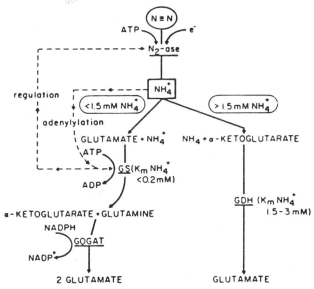

normal 20% have been found to increase N_2 fixation, but levels higher than 50% result in inhibition. The supply of available carbon and the supply of O_2 both have major effects on symbiotic fixation. Whether carbon or O_2 is limiting has been found to affect the efficiency not only of fixation but also of H_2 reutilization. The amount of ATP generated from a given quantity of O_2 appears greater when carbohydrate is oxidized then when H_2 is oxidized.

Ammonium is excreted by the bacteroids in nodules and also by cyanobacter associations. Free-living organotropic diazotrophs have been thought to produce NH_4^+ only for their own growth. In this case, death and decomposition of the diazotroph would have to occur before the plants obtained the fixed nitrogen. However, nitrogen-15 experiments with *Azospirillum* and the host plant, sorghum, indicate that the plant leaves contained 40% of the $^{15}N_2$ fixed over a 24-hr period. Thus at times there may be a rapid release of the NH_4^+ fixed in plant–microbe associations.

The NH_4^+ produced by bacteroids is incorporated into organic forms in the associated plant cells. Two methods of NH_4^+ incorporation are present in the nodules of temperate plants, glutamate dehydrogenase and the GOGAT enzyme system. The glutamate dehydrogenase pathway re-

Figure 10.11. Molecular structure of some nitrogenous solutes found in xylem sap. (Adapted from Atkins, 1982.)

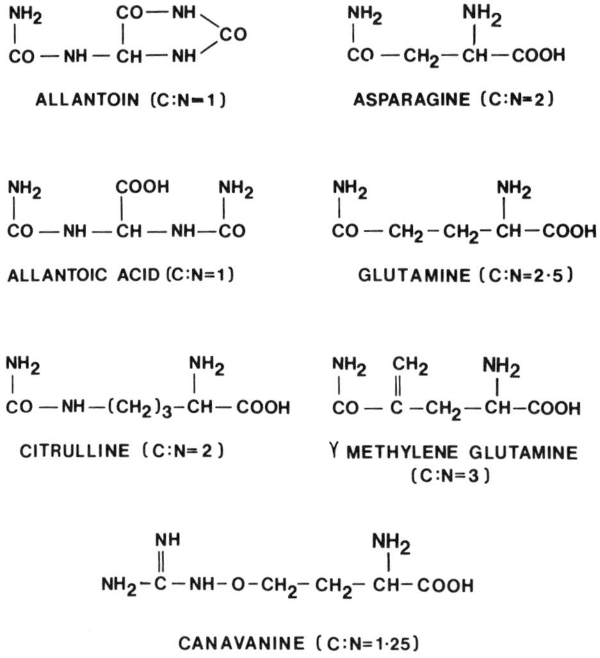

$$
\begin{array}{c}
NH_2 \quad\quad CO-NH \\
| \quad\quad\quad | \quad\quad\quad CO \\
CO-NH-CH-NH
\end{array}
$$

ALLANTOIN (C:N=1)

$$
\begin{array}{c}
NH_2 \quad\quad NH_2 \\
| \quad\quad\quad | \\
CO-CH_2-CH-COOH
\end{array}
$$

ASPARAGINE (C:N=2)

$$
\begin{array}{c}
NH_2 \quad COOH \quad NH_2 \\
| \quad\quad | \quad\quad | \\
CO-NH-CH-NH-CO
\end{array}
$$

ALLANTOIC ACID (C:N=1)

$$
\begin{array}{c}
NH_2 \quad\quad\quad NH_2 \\
| \quad\quad\quad\quad | \\
CO-CH_2-CH_2-CH-COOH
\end{array}
$$

GLUTAMINE (C:N=2·5)

$$
\begin{array}{c}
NH_2 \quad\quad\quad NH_2 \\
| \quad\quad\quad\quad | \\
CO-NH-(CH_2)_3-CH-COOH
\end{array}
$$

CITRULLINE (C:N=2)

$$
\begin{array}{c}
NH_2 \quad CH_2 \quad NH_2 \\
| \quad\quad \| \quad\quad | \\
CO-C-CH_2-CH-COOH
\end{array}
$$

γ METHYLENE GLUTAMINE
(C:N=3)

$$
\begin{array}{c}
NH \quad\quad\quad\quad\quad\quad NH_2 \\
\| \quad\quad\quad\quad\quad\quad\quad | \\
NH_2-C-NH-O-CH_2-CH_2-CH-COOH
\end{array}
$$

CANAVANINE (C:N=1·25)

quires more than 1.5 mM NH_4^+ in solution. Such high concentrations of NH_4^+ depress synthesis of nitrogenase. Use of the alternate NH_4^+ incorporation route, namely, the combination of glutamime synthetase (GS) and glutamate synthase (GOGAT), maintains the ammonium concentration at low levels. Glutamine synthetase also acts in the regulation of nitrogenase synthesis and in the production of bacteroid protein required for enzyme formation (Fig. 10.10).

Glutamine usually is transformed in secondary reactions to other products before leaving the root via the xylem. Many of the legumes growing in temperate climates transfer their nitrogen to the amides asparagine and glutamine; some tropical legumes export the ureides allantoin and allantoic acid. These nitrogenous compounds (Fig. 10.11) have lower C:N ratios and thus require lower amounts of carbon for reexport. Actinorhizal associations, such as the *Frankia–Alnus* symbiosis, also export the ureide citrulline. The amides asparagine and glutamine are more soluble than the ureides and thus probably are better adapted to the lower-temperature conditions under which temperate legumes exist. Other factors controlling whether ureides or amides are transported probably also exist, and the area requires further research.

Energetics of Nitrogen Fixation

Atkins and Rainbird (1982) quoted research data that show that on a theoretical basis the reduction of 1 mol of N_2 should require the energy present in 0.22 mol of glucose. Measurements of carbon utilization by N_2 fixation in both legumes and free-living bacteria result in differing ratios of H_2 to NH_4^+ production and varying levels of hydrogenase activity. Measured efficiencies of utilization indicate that a range of 0.66 to 1.38 mol of glucose are utilized per mole of N_2 fixed (Table 10.9). The cost of NH_4^+ assimilation into organics has been estimated at approximately 0.15 mol of glucose per mole of N_2. The transport of the fixed nitrogen as well as the growth of the nodule costs an additional 0.2–0.7 mol of glucose. This results in an overall range of 1.13 to 2.37 mol of glucose utilized per mole of N_2 fixed. Converted to a gram-carbon-per-gram-nitrogen basis, this results in an estimation of 2.9 to 6.1 g of photosynthetic carbon utilized per gram of nitrogen fixed in legume symbiosis. These are real costs that must be related to alternate costs of obtaining fixed nitrogen from the soil or from fossil fuels, when nitrogen is applied as a fertilizer. However, there are mitigating circumstances. The reduction of NO_3^- to NH_4^+ also requires extensive energy; the utilization of soil or fertilizer nitrogen present as NO_3^- should result in an energy consumption that is not greatly different from that required when the plant is fixing atmospheric N_2.

The nutrient uptake costs must be related to the overall ability of the

Table 10.9
Theoretically Based Cost Estimates for Energy Required by Component Processes in Actively Dinitrogen-Fixing Legume Nodules[a]

Item of functioning	Moles glucose per mole N_2 fixed	Grams carbon per gram nitrogen fixed
Nitrogenase–hydrogenase	0.66–1.38	1.7–3.5
Ammonia assimilation and related carbon metabolism	0.14–0.16	0.36–0.41
Transport of fixed nitrogen	0.13	0.33
Growth and maintenance of nodule	0.2–0.7	0.5–1.8
	1.13–2.37	2.9–6.1

[a]From Atkins and Rainbird (1982).

plants to grow under field conditions. The reduction of nitrate nitrogen by nitrate reductase in photosynthesizing leaves rather than in the root would affect the efficiency of NO_3^- utilization because of the greater availability of reductant in the leaves. In addition, plants with microbial symbionts have been shown to alter their leaf structure and efficiency of CO_2 uptake. This results in an increased efficiency of photosynthesis. The formation of symbiotic associations depends on a plant's ability to utilize otherwise underutilized photosynthetic capacity. The possibilities of plants compensating in photosynthesis for the needs of their symbiotic partners are discussed in Chapter 11.

Inoculation

The problems involved in attempting to introduce a new strain of *Rhizobium* or a genetically engineered organism into the soil environment are similar. This is related to the biological and environmental implications of inoculation of new organisms into soil. Much is already known about the use of rhizobia as inocula, and this should provide a model system for the use of genetically engineered organisms.

Extensive field experience has shown that there are certain qualities of an effective inoculant. Effective inoculants of *Rhizobium* may consist of a single strain or multiple strains. A unistrain inoculum is desirable where field tests have shown that a particular strain works best on a particular host under prevailing soil and climatic conditions and where growers generally plant that particular cultivar. A multistrain inoculant is preferable in areas where many varieties or cultivars may be grown and where there are wide variations in soil and climate. The multistrain inoculant, however, should not contain strains of rhizobia that form nodules without benefit to their host; such strains can be parasitic and impede nodulation by effective bacteria.

Field experience has shown that for maximum effectiveness the inoculant should provide at least 10,000 viable rhizobia on a small seed, such as alfalfa, and up to 1,000,000 rhizobia on a large seed, such as soybean. An average of 4 g inoculum kg^{-1} of seed is required for many seed types. The carrier medium should protect the rhizobia in the package for up to 6 months of shelf storage; this requires exchange of gases and retention of moisture. It should be easy to apply to the seed, adhere well, and protect the bacteria against drying, chemical fertilizers, or pesticides. Finally, the inoculum should be free of other bacteria, which might be detrimental to the rhizobia or to the seedling.

Legume inoculants are of two types: those designed for seed application and those designed for soil application. Seed inoculants are easiest to apply and are generally effective. High-quality inoculants carried on a peat substrate are generally considered to provide the most dependable inoculum. Methods of addition to seeds include the slurry technique, in which the inoculant is mixed with water containing a gum or sugar to improve adhesion. Mixing the inoculant without water or other liquid generally results in a less-uniform mixture of seed and bacteria.

Pelleting seeds with inocula has been found to be advantageous. Pelleting involves wetting the seeds with a peat-based inoculant slurry and then coating them with a compound such as finely pulverized limestone. Under tropical conditions, powdered rock phosphate is often substituted for limestone because limestone can be harmful to acid-tolerant bacteria. When a very large *Rhizobium* innoculum is required for hot, dry soils, or for soils heavily infested with ineffective native *Rhizobium,* an initial inoculant of 4 g inoculum kg^{-1} of seed is applied with a gum arabic slurry. Another 8 g kg^{-1} is mixed with the wet seed immediately before planting.

Water-soluble adhesives, such as polyvinyl acetate, polyurethane, polyurea varnishes, or resins dissolved in solvents, are used to bind limestone, rock phosphate, or peat-based inoculant to seeds. Advantages claimed for coated seed are more uniform seeding, easier planting, and better germination. Preinoculation of seeds by seed-processing companies months in advance of planting is attractive to farmers, but the *Rhizobium* death rate has been found to be rapid, and results with preinoculated seeds have not been consistent.

The rhizobia can be placed in moist soil below the seeds when the seeds are planted in hot, dry surface soils or when seeds are coated with toxic pesticides. Soil placement can introduce a larger inoculum of effective *Rhizobium* than that when applied directly to the seed in soil heavily infested with large populations of ineffective native bacteria; it, however, is more labor intensive and costlier. Experience, such as that with soybeans, has shown that it is very difficult to displace native strains with new strains. It is thus very important to ensure that areas being exposed to legumes for the first time are inoculated with the most effective strains available.

Challenges in Biotechnology

A number of genetic engineering studies involving N$_2$-fixing organisms are in progress or under consideration. The 17 bacterial genes responsible for N$_2$ fixation have been determined, and it is known that the genes tend to move as one group in transfers from organism to organism. However, much more than transfer of nitrogenase production is involved. Can rhizobia be structured to fix N_2 in the stems and leaves, to fix N_2 in the presence of plant-available nitrogen in the soil, to be successful competitors with other rhizobia and other organisms endemic in the soil, to overcome environmental stresses such as drought, acidity, and aluminum or other toxicities, and especially, to form successful symbioses with new host plants, even with nonlegumes? Symbiotic N_2 fixation is an expensive process relative to the energy required from the host plant. Increases in the efficiency of fixation involve better utilization of photosynthate and of hydrogenase in recapturing nitrogen. Successful nodulation of leaves or stems would dispense with the need for translocation of photosynthate below ground, would spare the inoculant from a multitude of competitors and antagonists in the soil, and would enable more precise definition of the roles of mycorrhizas and rhizobial symbioses by separating their arenas of activity.

The transfer of the complex oxygen-sensitive system to the leaf, such as in maize, however, will entail many problems. We need also to consider other approaches. It may be easier to alter an efficient N_2-fixing symbiotic legume genetically such that it has the growth potential and food characteristics of maize than to introduce N_2 fixation into maize.

The extensive molecular biological work on *Rhizobium* has developed the molecular linkage maps shown in Fig. 10.12. These include genes for attachment, root-hair-curling nodulation, host specificity, and nitrogenase. Empty boxes in the figure show areas that still require delineation. It may be found that the plant often is a limiting factor in symbiotic systems through its reactions to stress, its carbon allocation during fixation periods, and its photosynthetic pathway. The extensive work on *Rhizobium* should, however, act as a model system for the investigations needed on other soil organisms if molecular biological techniques are truly going to pave the way for major advances in rhizosphere dynamics.

It has been demonstrated that the ability of a number of rhizobia to interact with plants (infectivity) is determined by plasmid-borne genes. Plasmids vary from the very small, where they only code for about 3 genes, to some very large ones found in soil bacteria that can code for 200 or more genes. Transfer of the plasmids among these microbial symbionts can be correlated with transfer of host range specificity. A plasmid of major interest to soil microbiologists and plant molecular biologists is that of *Agrobacterium tumefaciens*. This close relative of *Rhizobium* causes tumors in plants by transferring some of its own genetic material to that of the host plant. This mechanism of plasmid transfer can be used to insert other genetic material into plant cells and is one of the few techniques available for transferring bacterial characteristics to plants.

Figure 10.12. Molecular linkage maps of the symbiotic region in *Rhizobium meliloti* and *R. leguminosarum*. *nif*, structural gene for nitrogenase; *fix*, genes necessary for dinitrogen fixation; *nod*, genes necessary for nodulation; *hac*, genes for root curling; *hsn*, genes for host specificity; *efn*, genes for nodulation efficiency; boxes with numbers, molecular weights of gene product; empty boxes, molecular weights yet to be established. (From Hodgson and Stacey, 1987.)

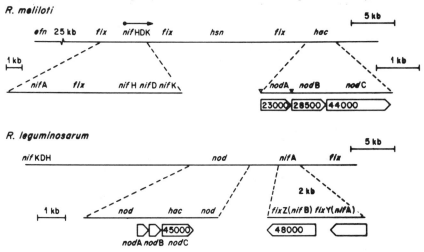

The transfer of the tumor-forming plasmid from *Agrobacterium* to *Rhizobium* was shown to convert the *Rhizobium* into a tumor-forming organism. The reverse transfer, that of the plasmids from *Rhizobium* to *Agrobacterium*, produced bacteria that induced nodulation but not N_2 fixation. This showed that N_2-fixing strains require either other genes on the chromosome or on other plasmids before the symbiotic relationship between the legume and the N_2-fixing symbiont can be established.

Plasmids are readily transferred among Gram-negative bacteria. The bacteria *Klebsiella*, *Rhizobium*, and *Azotobacter* can exchange DNA with *Escherichia coli*, the best-studied organism in genetic engineering. In the laboratory, drug-resistant strains for use in identification of organisms reintroduced into soil have been produced on fairly broad scales. These can be identified by their ability to grow on an antibiotic medium that inhibits the growth of most soil organisms.

The transfer of DNA from *Rhizobium* to *Azotobacter* strains not containing the necessary N_2-fixing genes resulted in the formation of a N_2-fixing strain. This showed that the N_2-fixing genetic material can be transferred as one unit.

References

Ahmadjian, V., and Paracer, S. (1986). "Symbioses: An Introduction to Biological Systems." University Press of New England, Hannover.

Atkins, C. A. (1982). Nitrogen fixation in nodulated plants other than legumes. *In* "Advances in Agricultural Microbiology" (N. S. Subba Rao, ed.), pp. 54–88. Butterworth, London.

Atkins, C. A., and Rainbird, R. M. (1982). Physiology and biochemistry in biological nitrogen fixation in legumes. *In* "Advances in Agricultural Microbiology" (N. S. Subba Rao, ed.), pp. 26–52. Butterworth, London.

Becking, J. H. (1978). Environmental role of nitrogen fixing blue-green algae and asymbiotic bacteria. *Ecol. Bull.* **26,** 266–281.

Becking, J. H. (1982). Nitrogen fixation in nodulated plants other than legumes. *In* "Advances in Agricultural Microbiology" (N. S. Subba Roa, ed.), pp. 90–110. Butterworth, London.

Burns, R. C., and Hardy, R. W. F. (1975). "Nitrogen Fixation in Bacteria and Higher Plants." Springer-Verlag, Berlin and New York.

Child, J. (1981). Biological nitrogen fixation. *In* "Soil Biochemistry" (E. A. Paul and J. N. Ladd, eds.), Vol. 5, pp. 247–322. Dekker, New York.

Dobereiner, J., and Day, J. M. (1975). Nitrogen fixation in the rhizosphere of tropical grass. *In* "Nitrogen Fixation by Free-Living Micro-Organisms" (W. D. P. Stewart, ed.), *Int. Biol. Programme* 6, pp. 39–56. Cambridge Univ. Press, London and New York.

Fogg, G. E. (1977). Nitrogen fixation. *In* "Algal Physiology and Biochemistry" (W. D. P. Stewart, ed.), pp. 560–582. Blackwell, Oxford.

Havelka, U. D., Boyle, M. G., and Hardy, R. W. F. (1982). Biological nitrogen fixation. *Agronomy* **22,** 365–413.

Hodgson, A. L., and Stacey, G. (1987). Potential for *Rhizobium* improvement. *CRC Crit. Rev. Biotechnol.* **4,** 1–73.

Millbank, J. W. (1977). Lower plant associations. *In* "A Treatise on Dinitrogen Fixation" (R. W. F. Hardy and W. S. Silver, eds.), Sect. 3, pp. 125–184. Wiley (Interscience), New York.

Patriquin, D. G. (1982). New developments in grass bacteria associations. *In* "Advances in Agricultural Microbiology" (N. S. Subba Rao, ed.), pp. 139–190. Butterworth, London.

Peters, G. A., and Calvert, H. E. (1982). The *Azolla–Anabaena* symbiosis. *In* "Advances in Agricultural Microbiology" (N. S. Subba Rao, ed.), pp. 191–218. Butterworth, London.

Postgate, J. R. (1982). "Fundamentals of Nitrogen Fixation." Cambridge Univ. Press, London and New York.

Shubert, K. R., and Evans, H. J. (1976). Hydrogen evolution. A major factor affecting the efficiency of nitrogen fixation in nodulated symbionts. *Proc. Natl. Acad. Sci. U.S.A.* **73,** 1207–1211.

Stewart, W. D. P. (1975). "Nitrogen Fixation by Free-Living Microorganisms." Cambridge Univ. Press, London and New York.

Stewart, W. D. P. (1977). Blue-green algae. *In* "A Treatise on Dinitrogen Fixation" (R. W. F. Hardy and W. S. Silver, eds.), Sect. 3, pp. 63–124. Wiley (Interscience), New York.

Tu, J. C. (1975). Rhizobial root nodules of soybeans as revealed by scanning and transmission electron microscopy. *Phytopathology* **65,** 447–454.

Supplemental Reading

Bauer, W. D. (1981). Infection of legumes by rhizobia. *Annu. Rev. Plant Physiol.* **32,** 407–409.

Beringer, E. R. (1982). Microbial genetics and biological nitrogen fixation. *In* "Advances in Agricultural Microbiology" (N. S. Subba Rao, ed.), pp. 26–52. Butterworth, London.

Dommergues, Y. R. (1982). Scarcely explored means of increasing the soil N pool through biological N_2 fixation. *Proc. Int. Congr. Soil Sci., 12th, 1982,* pp. 138–149.

Granhall, U. (1978). Environmental role of nitrogen fixing blue-green algae and asymbiotic bacteria. *Ecol. Bull.* (Stockholm) **26.**

Long, S. R. (1984). Genetics of rhizobium nodulation. *In* "Plant Microbe Interactions" (E. Nester and G. Kosuge, eds.), pp. 265–306. Macmillan, New York.

Parker, C. A. (1977). Prospectives in biological dinitrogen fixation. *In* "A Treatise on Dinitrogen Fixation" (R. W. F. Hardy and W. S. Silver, eds.), Sect. 3, pp. 367–471. Wiley (Interscience), New York.

Pate, J. S. (1977). Functional biology of dinitrogen fixation by legumes. *In* "A Treatise on Dinitrogen Fixation" (R. W. F. Hardy and W. S. Silver, eds.), Sect. 3, pp. 473–517. Wiley (Interscience), New York.

Shanmugan, K. T., Gara, F. O., Anderson, K., Morandi, C., and Ballantyne, R. C. (1978). Control of biological nitrogen fixation. *In* "Nitrogen in the Environment" (D. R. Nielsen and J. G. MacDonald, eds.), Vol. 2, pp. 393–416. Academic Press, New York.

Vincent, J. M. (1977). Rhizobium general microbiology. *In* "A Treatise on Dinitrogen Fixation" (R. W. F. Hardy and W. S. Silver, eds.), Sect. 3, pp. 277–368. Wiley (Interscience), New York.

Food and Agricultural Organization of the United Nations (1984). "Legume Inoculants and Their Use." FAO, Rome.

Chapter 11

Mycorrhizal Relationships

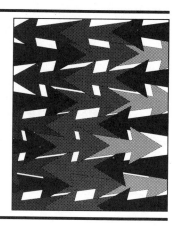

Introduction

Mycorrhizas are part of the array of symbiotic associations between microorganisms and plants (Table 11.1). In the case of the N_2-fixing plants the relationships is tripartate. In legumes this consists of the mycorrhizal fungus, the host plant, and *Rhizobium*. In some shrubby nonlegumes, the association consists of the green plant, the mycorrhizal fungus, and the actinomycete *Frankia*. The recognition of the role of all the partners is important. The study of only two of the components at any one time can lead to erroneous conclusions concerning the activities of a symbiosis in the field.

Fossil records show that the fungus–plant root association recognized as a mycorrhiza is as old as terrestrial plants. The earliest known plants, the fossils of the Rhynie chert of 370 million years ago, have identifiable fungal structures very similar to present-day vesicular–arbuscular mycorrhizas, as described in this chapter. It is of interest that in the fossil chert association the underground plant organs with which the fungi were associated were not true roots. Over time, there has been more adaptation in the plant than in the fungus.

In their review of the history of this field, Trappe and Berch (1985) stated that the first illustration of a mycorrhiza was published in 1840, when Robert Hartig illustrated the mycorrhizal fine roots of pine. However, he did not recognize the fungal component as a separate entity. The fungal cells associated with orchids were described and recognized as fungi by S. Reissek in 1847. In 1881, F. Kamienski pointed out that certain tree roots had a fungus layer surrounding them and that nutrients must penetrate this fungus layer to be taken up by the plant. A. B. Frank, in 1885,

Table 11.1
Major Microbial Symbiotic Associations That Occur in Nature

Association	Symbionts
Lichens	Fungi and algae
Lichens	Fungi and cyanobacteria
Corals	Coelenterates and algae
Legumes	Legumes, *Rhizobium*, and vesicular–arbuscular mycorrhizas
Water ferns	*Azolla* and cyanobacteria
Actinorhizas	shrub and tree angiosperms, *Frankia* and vesicular–arbuscular mycorrhizas
Mycorrhizas	Most plants and fungi

asked to study the possible culture of truffles in Prussia, recognized the fungus–root structure and coined the name *Mykorrhizen* to indicate fungus root. He characterized the sheath-forming fungus on tree roots as *ecto-tropisch* (ectotrophic). The association without the compact external sheath but with hyphal penetration within the cells of roots was described as *endotropisch* (endotrophic).

Forms and Distribution of Mycorrhizas

Mycorrhizal fungi are associated with nearly all plants. Certain species of the families Cyperaceae, Juncaceae, Urticaceae, Chenopodiaceae, Caryophyllaceae, and Brassicaceae (Crucifereae) are less commonly infected than others. It is of interest that some of the more obnoxious weeds in cultivated agriculture are nonmycorrhizal. They depend on a fast growth habit and uninfected fine root hairs to develop rapidly in areas where nutrients are generally adequate (Trappe, 1987).

There is a general negative relationship between the occurrence of fine roots or root hairs and mycorrhizal formation. Species with fine roots and many root hairs, such as the grasses, are not nearly as dependent on their fungal partners for nutrition as are the tap-rooted species of legumes or coarse-rooted plants, such as orange trees and pines. Those plants that form mycorrhizas have one common attribute—they can control the potential pathogenicity of the fungi. The fungus *Rhizoctonia* is a potent pathogen of maize but forms stable mycorrhizas with orchids. Except for the fungi associated with orchids, mycorrhizal fungi generally are considered to have a weak ability to compete in nature for organic forms of carbon with other decomposing organisms. They depend on the host for their carbon supply. Forms of the Basidiomycete *Boletus*, capable of growth on plant residues and soil organic matter (SOM), have active cellulases.

Table 11.2

Characteristics of the Important Kinds of Mycorrhiza[a]

Characteristic	Vesicular-arbuscular	Ectomycorrhiza	Ectendomycorrhiza	Arbutoid	Monotropoid	Ericoid	Orchid
Fungi							
Septate	−	+	+	+	+	+	+
Aseptate	+	+	−	−	−	−	−
Hyphae enter cells	+	−	+	+	+	+	+
Fungal sheath present	−	+	+ or −	+	+	−	−
Hartig net formed	−	+	+	+	+	−	−
Hyphal coils in cells	+	−	+	+	−	+	+
Haustoria							
Dichotomous	+	−	−	−	−	−	−
Not dichotomous	−	−	−	−	+	−	+ or −
Vesicles in cells or tissues	+ (or −)	−	−	− (or +)	+	−	−
Achlorophylly	− (or +)	−	−	−	+	−	+
Fungal taxon[b]	Phyco	Basidio Asco Phyco	Basidio Asco	Basidio	Basidio	Asco (Basidio)	Basidio
Host taxon	Bryophyte Pteridophyte Gymnosperm Angiosperm	Gymnosperm Angiosperm	Gymnosperm Angiosperm	Ericales	Monotropaceae	Ericales	Orchidaceae

[a]The structural characters given relate to the mature state, not the developing or senescent states. Entries in brackets, rare.

[b]Prefixes for -mycete.

Related species capable of forming mycorrhizas do not have cellulitic capability. Several major ectomycorrhizal fungi have been shown to be capable of degrading plant lignin and lignocellulose. Their role in nature is not yet known.

The original ectotrophic and endotrophic classification of Frank has been expanded as more information on the plants and their symbiotic partners has become available. Table 11.2 describes the seven forms recognized by Harley and Smith (1983). Frank's ectotrophic classification has continued to this day, but the original endotrophic description has been subdivided on the basis of more information. Groups such as the orchids, monotropoids, and ericoids represent associations with specific host types. The ectendotrophic association demonstrates the continuum that exists between vesicular–arbuscular mycorrhizas and ectomycorrhizas.

Vesicular–Arbuscular Mycorrhizas

Vesicular–Arbuscular Mycorrhizas (VAM) the most common of mycorrhizal forms, involve fungi classified as Zygomycetes. The aseptate hyphae enter root cells of nearly all cultivated plants and of many forest and shade trees, shrubs, and wild herbaceous species. No discernible root or outside structural changes are noticeable. On some species, such as onion, there is a slight yellowing of the roots, but most other plants have to be examined under the microscope to determine the presence of VAM. The vesicular–arbuscular name is derived from the internal vesicles used for storage. These 10 to 100-μm expansions of the fungal hyphae between cells (Fig. 11.1) are usually filled with lipids. The other internal structure in the root cortex is the arbuscule. This consists of finely branched hyphae similar to the haustoria of plant pathogens (Fig. 11.2). The arbuscules persist within individual plant cells for a 4- to 10-day period. After this time they are digested by the plant cell, and new ones are formed in other cells. Nutrient transfer is thought to occur between the finely branched fungal mycelium and highly invaginated plant cell membranes. It has been observed that the weight of plant cytoplasmic material within an arbuscular plant cell is 20 times that of an uninfected cell.

External structures of VAM fungi are hyphae that penetrate the soil, and individual resting spores. The latter are produced asexually on straight, subtended hyphae and are known as chlamydospores. Some VAM fungi have an aggregation of spores in a sporocarp. Structures on the exterior of plant roots that have in the past been identified as vesicles are now considered to be chlamydospores in the formative stage. The term vesicle should be restricted to the expanded hyphal tips within the plant cortex.

Five major genera of the family Endogonaceae form VAM. These in-

Figure 11.1. Schematic diagram of the association of vesicular–arbuscular mycorrhizal fungi and a plant root. The external mycelium bears large chlamydospores (CH) and occasional septate side branches (SB). Infection of the plant can occur through root hairs or between epidermal cells. Arbusculae at progressive stages in development and senescence are shown (A–F) as is a vesicle (V). To avoid confusion, cell walls of the root are not indicated where they underlie fungal hyphae. From a drawing by F. E. Sanders, p. 129, quoted in "Plant Root Systems" by R. Scott Russell (1977). Copyright the McGraw Hill Book Company.)

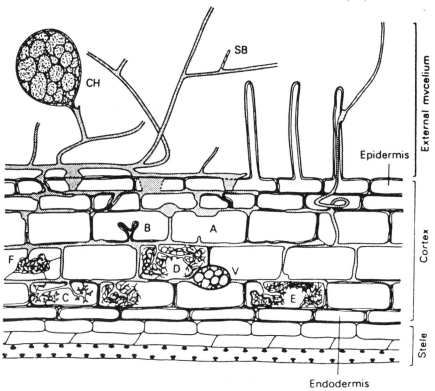

clude *Glomus*, *Gigaspora*, *Acaulospora*, *Sclerocystis*, and *Scutellospora*. They are distinguished by the morphology of their resting spores (Figs. 11.3 and 11.4). The genus *Glomus* contains a number of species that have individual thick-walled terminal chlamydospores. *Glomus* also produces sporocarps that are multiple sporing structures having individual chlamydospores within them. At maturity, spores separate from the hypha.

The genus *Gigaspora* rarely forms vesicles but has spores budded from a bulbous, suspensor-like tip of a hypha. The spores have a resemblance to zygospores, the sexual stages of Zygomycetes. Because *Gigaspora* is asexual, its spore has hitherto been termed an azygospore, but many my-

Figure 11.2. Arbuscule development, showing how the hyphal branches fill the cytoplasmic volume. (From Kinden and Brown, 1976.)

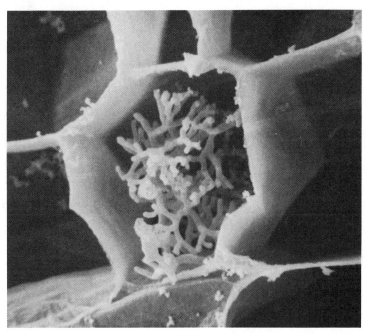

cologists refer to it simply as a spore. The genus *Acaulospora* also produces resting terminal spores. These are formed from a previously developed but short-lived mother spore. The spores are found singly in soil or sometimes within roots. The genus *Sclerocystis* contains a number of chlamydospores arranged around a central mass of interwoven hyphae that form a sporocarp (Fig. 11.4).

Spore germination results from the regrowth of attached hyphae in the case of *Glomus* and *Sclerocystis,* and through the wall of the spore in *Gigaspora* and *Acaulospora.* If the growing hypha of a germinating spore does not contact a susceptible root, it can restrict its growth; the spore then continues in a resting phase. Most mycologists believe that the direction of hyphal growth from the spore is random. Generally, there appears to be no message from plant roots to germinating spores to ascertain the direction of the growing hyphal tip until the hypha is within a few millimeters from the root. The hypha then produces a fan-shaped complex of narrow (2 to 7 μm) branches that produce infection in the root.

Figure 11.3.   Cross-section of three representative vesicular arbuscular mycorrhizal spores. A, *Gigaspora heterogama* and subtending hypha, spore size 160 μm; B, *Glomus invermaius* and subtending hypha, spore size, 75 μm; C, *Acaulospora trappei* budded from tapered stem of parent cell, spore size 80 μm. (From Trappe and Schenk, 1982.)

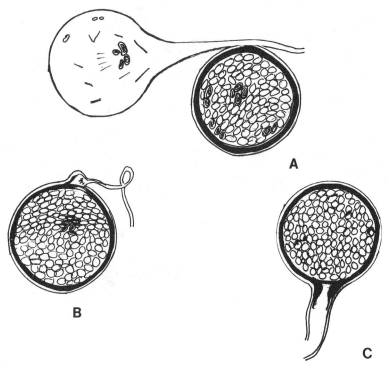

Individual VAM spores are from 10 to 400 μm in diameter, but mostly are in the 40- to 200-μm range. Identification requires a number of spores for microscopic observation by a person who is familiar with spore size, structure, and wall decoration. It also requires crushing to observe the wall structures. A fine endophyte with hyphae of 1 to 3 μm in diameter rather than the 5 to 10 μm of other VAM fungi is characterized by small spores of approximately 10 μm in diameter. *Glomus tenuis*, a fine endophyte, is said to be particularly effective in phosphorus uptake under fertile soil conditions. However, Harley and Smith (1983) stated that the classification of this endophyte is still open to question and that it should probably be referred to simply as a fine endophyte until further characterization is achieved.

Ectomycorrhizas

The ectomycorrhizas (ECM) (Table 11.2) consist of septate fungal cells infecting the roots of temperate trees and shrubs. The fungi form a compact mantle or sheath over the root surface and penetrate between the cells of the root cortex to form a complex intercellular system called the Hartig net (Fig. 11.5) (Harley and Smith, 1983). The mantle on the exterior can vary in thickness, and the Hartig net between the cells is also more obvious in some mycorrhizal associations than in others. The typical forked branching shown in Fig. 11.6 is characteristic of ECM in pines, whereas the rootlets of *Larix* (larch) and *Tsuga* (hemlock) do not have the extensive branching of the root tips but deeper, more complex mantles. Oak trees often show no root branching attributable to ECM infection.

Ectomycorrhizal fungi produce auxins that are responsible for some of the morphological differences between mycorrhizal and nonmycorrhizal roots. The short roots are thought to precede rather than be the result of infection. Many fungal species form ECM. Basidiomycetes, Ascomycetes, and Phycomycetes are involved. Many mushrooms, including representatives of the Agaricaceae, Boletaceae, Russulaceae, and Cortinariaceae induce extensive mycorrhizas. They also produce known edible above ground and underground fruiting bodies. Some ECM are specialized and are associated with only a single host species; others, such as *Cenococcum graniforme* (an imperfect ascoymcete), have a wide host range.

The effects of ECM fungi have been well documented. It is usually difficult, and sometimes impossible, to plant seedlings of trees in grassland soils or other new areas without introduction of the mycorrhizal partner. A number of the significant ECM fungi can be grown away from the host plant, thus making experimentation and the possibility of inoculum production possible. Inoculation with mycorrhizal partners, such as *Pisolithus tinctorius* and *Telephora terrestrius,* can greatly enhance the growth of pine, especially on sites, such as mine spoils, with poor nutrition and toxic metals. These fungi are thought to increase absorption of nutrients and to produce antibiotics. They also produce a barrier to root pathogens, such as *Pythium, Rhizoctonia,* and *Phytophthora.*

The external mantle may be much reduced or even absent in some mycorrhizas on trees, thus forming ectendomycorrhizas. The Hartig net is well developed, but the hyphae also penetrate into the cells of the host. The same fungus that produces ECM may, on a different host plant or under different conditions, form ectendomycorrhizas, which have some characteristics of both ECM and VAM.

There is a general concentration of ECM in forests of north and south temperate and subarctic regions. They also occur at high elevations in the

Figure 11.4. Scanning electron micrographs of *Glomus* spp. (*A*) Chlamydospores (C) and hyphae (P) of *G. fosciculatus* on a soybean root; (*B*) Sporocarp (S), probably containing a single spore and chlamydospore (C) of *G. mosseae;* (*C*) Chlamydospore of *G. mosseae*, showing the funnel-shaped subtending hypha (arrow). (From Brown and King, 1982.)

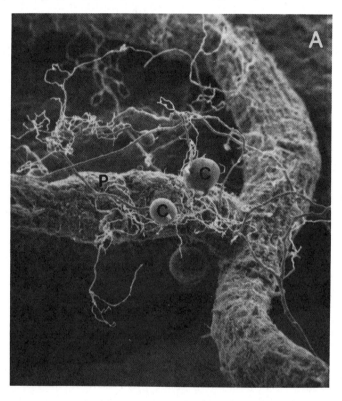

tropics. At one time it was considered that the tropical rain forests contained strictly VAM; further research has shown ECM in some of the trees and in even some tree legumes. Species of the genera *Alnus, Cupressus, Juniperus, Acacia,* and *Casuarina* have been found to have both ecto- and ectendoymcorrhizas.

Ericaceous Mycorrhizas

Harley and Smith have separately classified the two types of mycorrhiza associated with the Ericales and those associated with the Monotropaceae (Table 11.2). The arbutoid mycorrhizas are formed as a mantle of a few to 85 μm in thickness. This, like the mantle in ECM, can serve as a storage organ. The Hartig net, when found, only penetrates the outer layer of

Figure 11.4. (*continued*)

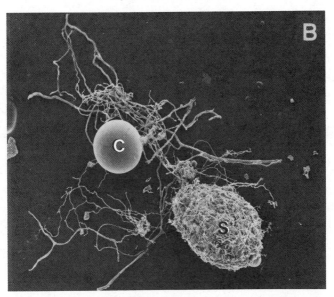

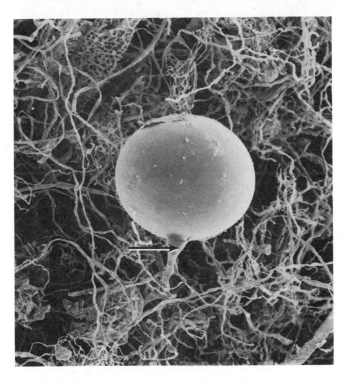

Figure 11.5. SEM of cross-section of pine root with mantle of ectomycorrhizal fungi. A, 30×; B, 300×.

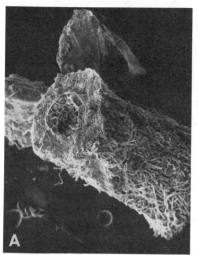 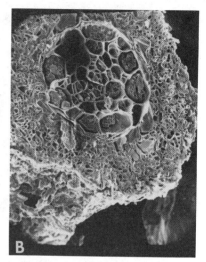

cortical cells. The septate hyphae form intracellular coils that eventually disintegrate within the cell. As there is little lateral spread within the cortex, each new penetration of a cell requires hyphal entry from the soil. The root system of *Arbutus* trees is differentiated into long and short roots, and only the short roots are extensively infected. Examples of the arbutoid type include the rhododendrons and *Arctostaphylos* (manzanita), found extensively in the Mediterranean-type climate of California.

Ericaceous plants are characteristic of acid or peaty soils. Hosts include *Calluna* (heather), and *Vaccinium* (blueberries). The heaths of western Europe, the subarctic tundra, and high-altitude moist heaths in South Africa, Austria, Pakistan, and Australia represent major examples of ericaceous mycorrhizas. Only one type of fungus, a dark, slow-growing septate Ascomycete, has been found associated with the fine roots of heath plants. This fungus, now known as *Pezizella*, forms multiple coils within the cells. Up to 42% of the volume of the root cells can be occupied by the fungal hyphae. Most mycorrhizal fungi are known to assist in the absorption of nutrients. Research with ericoid mycorrhizas has shown an increased uptake of soil organic nitrogen in the presence of $^{15}NH_4^+$, demonstrating a possible degradative capacity of these fungi.

The achlorophyllous, herbaceous plants in the Monotropaceae were originally thought to be saprophytes. It is now considered that a basidiomycete *(Boletus)* infects both the *Monotropa* and the roots of neighboring

Figure 11.6. The club-shaped forked branching of pine mycorrhiza, 20×.

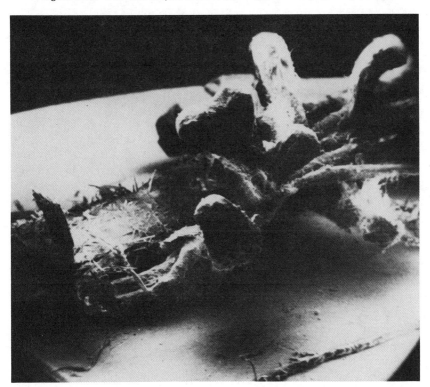

trees. The very small seed of *Monotropa* will initiate germination but will not develop further until infected with *Boletus*. [^{14}C] Glucose and ortho [^{32}P] phosphate injected into spruce and pine trees were translocated by the fungi over a distance of 1 to 2 m to the tissues of *Monotropa*. The advantage to *Monotropa* is clear; the advantage to the fungus is not so clear, and the host tree must be looked on as being parasitized.

Mycorrhizas of Orchidaceae

The septate fungi forming a symbiosis with the 12,000 to 30,000 species of the orchid family merit separate consideration because of the physiology of the association and the interest that orchids generate in the horticultural literature. Orchid seedlings are very small, with the largest being approximately 14 μg in weight. All orchids, whether chlorophyllous or achlorophyllous as adults, pass through a seedling stage during which they are unable to photosynthesize. Since orchid seeds are too small to contain reasonable reserves, a germinating embryo does not develop further unless it receives an outside supply of carbohydrates or is infected by a compatible mycorrhizal fungus. At one time these fungi were assigned to the imperfect, clamp-forming *Rhizoctonia*. A number of forms have now been induced to produce fruiting bodies. Genera such as *Marasmius, Armillaria,* and *Fomes* have been found represented. In their nonmycorrhizal stages these genera are often destructive pathogens.

The infection in orchids spreads from cell to cell, with hyphal coils taking up a large portion of the volume of infected cells. Intracellular hyphal coils have a limited life even in stable mycorrhizal associations of orchids. Hyphal degeneration can occur as early as 30 to 40 hr after initiation of infection and is usually complete within 11 days. However, reinfection of the same cell or hyphal penetration from a neighboring cell can take place. Digestion of the fungus was at first thought to be the basis for nutrient transfer. It is now believed that nutrient transfer is a continuous process more likely to occur across the intact membranes of the fungus and the closely associated host cells. There is very little direct evidence that as the orchid matures and becomes photosynthetic that a net transfer of carbon from the host to the fungus takes place.

Little work has been done on the mycorrhizal association between the adult orchid or with the green orchids existing in temperate forests. Tropical epiphytic orchids have hyphal connections penetrating the tissue of supporting plants, indicating that parasitism of the host plants can occur, as described earlier for *Monotropa*. However, the fungus in *Monotropa* possesses a well-developed fungal sheath, a Hartig net, and specialized haustoria, all of which orchid associations lack.

Physiology and Function of Mycorrhizas

The same general physiological processes apply to most of the wide range of mycorrhizal associations occurring in nature. The major contribution of the fungus is in nutrient uptake and translocation, especially that of phosphorus. Mycorrhizal plants also have been said to be able to withstand drought better. Other effects that are known to occur are protection from plant pathogens, nematodes, and heavy-metal concentrations in the rooting zone. The fungi in turn have a requirement for soluble plant carbon. The balance between the ability of the plant to provide otherwise underutilized photosynthetic capacity to supply the fungi with carbon while the plant obtains nutrients is the basis of a successful symbiosis. Parasitism can occur when the nutrient levels in the soil are such that the fungi cannot extract extra nutrients, and no benefit accrues to the plant in return for the carbon transferred to the fungus.

Nutrient Uptake

The nutrients phosphorus, nitrogen, zinc, copper, and sulfur have been shown to be absorbed and translocated to the host by mycorrhizal fungi. Extensive work has shown that the mechanism of uptake is the greatly increased soil volume explored by the mycorrhizal fungus. In addition, the fungus appears to be able to absorb phosphorus at lower solution concentrations than an uninfected plant root. Mycorrhizal fungi have active phosphatase activity. Whether this is primarily for the use of the fungi in the internal translocation of the phosphorus or is an uptake mechanism is not known. Experiments have shown that phytate, an organic phosphorus source, is decomposed by mycorrhizal fungi, and that the phytate phosphorus is taken up. Other experiments have shown that mycorrhizas have no greater access to organic phosphorus than nonmycorrhizal roots.

The diffusivity of the NO_3^- ion in soil is 10^{-6} cm^2 sec^{-1}, that of NH_4^+ is 10^{-7} cm^2 sec^{-1}, and that of PO_4^{3-} is 10^{-8} cm^2 sec^{-1}. The NO_3^- ion is present in sufficiently high concentrations in many agricultural soils and has such a high diffusion coefficient that the mycorrhizas have little effect on nitrogen nutrition. The largest effect on nutrient uptake is found for phosphorus. Droughty soils have 10 to 100 times lower diffusion rates than wet soils. In such soils, phosphorus uptake is often limited without the mycorrhizal fungi. Generally, positive plant growth responses to mycorrhizal infections are found in soils in which the concentration of a nutrient such as phosphorus is in low concentration in the aqueous phase, but the soil has a reserve of adsorbed or other solid forms. The lowest soil solution concentrations of phosphorus at which various field crops start benefiting from mycorrhizal association ranges from 0.1 μg ml^{-1} for soybeans to 1.6 μg ml^{-1} for cassava and *Stylosanthes*. The latter two crops require mycorrhizal infection for growth under field conditions.

Citrus has a coarse root structure and usually responds strongly to mycorrhizal infection. Figure 11.7 shows that mycorrhizal sour orange seedlings responded favorably to mycorrhizal fungi at all phosphorus levels, whereas Troyer orange showed favorable response only at low phosphorus levels. The data shown in Fig. 11.8 for soybeans is typical of that obtained for many plants. Inoculation with VAM fungi resulted in 30 to 60% of the yield potential possible with fertilizer phosphorus addition. Most plants have the capability of inhibiting mycorrhizal infection at adequate concentrations of phosphorus and nitrogen. However, experimentation has shown that it is possible to isolate fungi that can infect at high soil phos-

Figure 11.7. Mean dry weights, per seedling, of mycorrhizal (——) and non-mycorrhizal (– – –) citrus seedlings (A, Brazilian sour orange; B, Troyer citrange) fertilized with different amounts of fertilizer phosphorus (P). (From Menge *et al.,* 1978.)

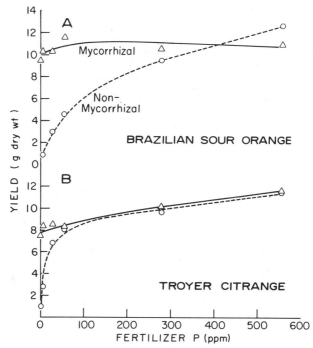

phorus levels. It should therefore be possible to increase the mycorrhizal effect at phosphorus levels further up the growth response curve.

Intensive agriculture can afford to utilize expensive sources of fertilizer phosphorus, but large areas of the world do not have deposits of phosphorus, nor can they afford to import it. Even in developed nations, extensive management systems, such as grasslands and forests, cannot be economically fertilized with phosphorus. Many soils, such as those of the savannahs of Africa and South America, have high capacity for phosphorus fixation. Mycorrhizal fungi appear beneficial in such situations. In addition to absorbing phosphorus, they may be able to intercept recently mineralized organic phosphorus before it is fixed in the mineral forms.

Plants that are susceptible to mycorrhizas usually are already infected with native strains. Responses to ECM infection are very easy to obtain in nurseries where sterilization has eliminated the native mycorrhizas. Other examples include transfer of ECM trees to new habitats where

Figure 11.8. Dry weight for plants inoculated with vesicular–arbascular mycorrhizal fungi [*Glomus fasciculatum* (∇) and *G. mosseae* (△)] or left uninoculated and fertilized with phosphorus (P) (•). Soybean shoot, $\hat{y} = 24.7 - 22.9^{-2.4x}$; soybean root, $\hat{y} = 6.8 - 5.7^{-2.5x}$. (Adapted from Pacovsky et al., 1986.)

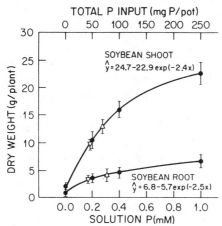

growth failed unless mycorrhizas were present. Experiments with VAM under field conditions usually are not as striking. However, a number of VAM field inoculation experiments, with both legumes and nonlegumes, have shown moderate increases in soils with low to moderate levels of available phosphorus and low levels of native mycorrhizas (Hayman, 1987).

The mechanism of phosphorus translocation within the mycorrhizas is believed to be due to cytoplasmic streaming. Polyphosphate makes up a significant portion of the phosphorus of mycelial strands, and it could be translocated via streaming. Polyphosphate also can act as a storage form for phosphorus, both in the mycorrhizal hyphae and in the sheath of ECM. Its insolubility maintains low internal phosphorus concentrations, thus allowing for its continued uptake from the soil solution.

Transfer of nutrients to the plant is across the arbuscule in VAM, through the coiled hypha of orchids and ericoids, and through the Hartig intercellular net of ECM. The close association of the fine fungal structures and the specially developed plasmalemmae of the plant cell is considered to provide the site of nutrient transfer. The sheath and Hartig net of the ECM have a persistence time of 9 to 14 months. The arbuscles of VAM and the finely coiled hyphae of the ericoids and orchids are continually being formed and reformed. A turnover time of 4 to 15 days is found for both types of internal fungal structures.

Nitrogen effects have not been as often noted as have those for phosphorus, but experimentation with nitrogen is not as straightforward as with phosphorus. It has been established that the uptake of NH_4^+ is facilitated by mycorrhizas, especially the ECM of forest trees. There are now two examples (one with ericacous mycorrhiza and one with VAM) of increased availability of otherwise nonutilized organic nitrogen with mycorrhizal fungi. Isotopic dilution experiments in the presence of $^{15}NH_4^+$ showed that a higher proportion of unlabeled nitrogen entered the roots of mycorrhizal plants. This indicated that the mycorrhizal plant had greater access to a nonfertilizer source of nitrogen than the nonmycorrhizal plant.

Only some of the mycorrhizal fungi (e.g., Basidiomycetes) appear to have nitrate reductase. It is a prerequisite for NO_3^- utilization. Ammonium is incorporated into organic compounds prior to transfer out of the root region. The tricarboxylic acid cycle intermediates required for biosynthesis of NH_4^+ into organic compounds are brought down to the root or are produced *in situ* by the fixation of CO_2 by the root phosphoenolpyruvate (PEP) carboxylase system. The mechanism of nitrogen incorporation into organic compounds in mycorrhizal fungi is similar to that previously discussed for nitrogen immobilization and N_2 fixation. Root $^{14}CO_2$ fixation by an organ known to evolve CO_2, although not significant from a total carbon-flow-balance basis, can affect studies that use tracer carbon as an index of transport and in calculations of energy requirement based on carbon-14 experiments.

Zinc and copper have been shown to be taken up by mycorrhizas; under deficient conditions there are increased plant yields. However, the occurrence of zinc and copper deficiencies is not very common, and the deficiency can easily be overcome by the application of low levels of fertilizer, either as a foliar spray or as a soil application.

Heavy-metal protection has been shown for plants growing on mine spoils in the presence of large concentrations of zinc, cadmium, and manganese. Protection from manganese toxicity in a VAM-infected legume *(Vicia faba)* growing on a high-manganese soil also has been demonstrated. It is thought that the heavy metals are bound by carboxyl groups in the pectic compounds (hemicelluloses) of the interfacial matrices, between the fungus and the host cells. It has also been demonstrated that plants growing on mine spoils with heavy-metal contamination have mycorrhizas selected for greater resistance to these metals.

Other Effects

Plants growing in arid environments are heavily mycorrhizal, and there are numerous statements about the increased drought resistance of mycorrhizal plants. Interpretation of such data is made difficult by the interactions of phosphorus with water usage. Increased phosphorus levels generally increase drought resistance. Calculations made for VAM fungi indicate that the amount of water that could travel through the mycelia to the plant is not large enough to influence plant growth or survival. However, work with ECM fungi showed that the fungal strands are capable of altering the water potential of plants. Seedlings were maintained in a healthy state for a 10-week period when the only source of water was through mycelial strands growing in moist peat. Seedlings lost color and died within 1 week of severing the mycelial strands.

The mycorrhizal association results in altered roots, especially in ECM, where the presence of the thick external sheath affords physical protection. Protection from pathogenic fungi, such as *Phytophthora, Pythium,* and *Rhizoctonia,* and from nematode attack has been demonstrated. The mechanism could be mechanical protection; antibiotic production could also play a role.

Mycorrhizal fungi have an aggregating effect, especially on coarse, sandy soils. Stabilization of dune sands and mine spoils is facilitated by plants with extensive mycorrhizal production.

Hormone production by the ECM is responsible for the club-shaped roots that are typical of this association. The interaction of hormones, phosphorus nutrition, and photosynthesis in VAM associations is more difficult to evaluate. Allen *et al.* (1980) have found higher cytokinin levels in leaves and roots of grass in a VAM association. They stated that cytokinin increases in plants are known to influence photosynthesis and transpiration rates, phosphorus uptake, and ion transport.

Carbon Flow in Mycorrhizal Plant Associations

The carbon flow from the host to the mycorrhizal fungus supplies the carbon required for growth of the fungus in all associations, with the exception of orchids. Whether ECM fungi capable of degrading cellulose and lignins utilize any of the carbon is not known. In VAM, the conversion of plant carbon to glycolipids constitutes a mechanism for the accumulation of substrate by the fungus. In ECM, the production of carbohydrates, such as trehalose and mannitol, that are not readily metabolized by the plant cells may have a similar function in providing a concentration gradient allowing sucrose to flow to the fungus.

The fraction of root weight attributable to mycorrhizal fungi approximates 3% in the fibrous VAM root system of sorghum and in the ECM of a 180-year-old *Abies* (silver fir) (Table 11.3). Values of 12 to 16% were found in VAM legumes and nonlegumes and in younger (23 years) ECM

Table 11.3
Biomass of Mycorrhizas in Roots of Symbiotic Associations

Fungus	Host	Root weigh (%)	Method	Infection (%)
Glomus fasciculatum	*Centrosema pubescens*	14	Chitin	95
Glomus mosseae	*Centrosema pubescens*	7	Chitin	100
Acaulospora laevis	*Centrosema pubescens*	5	Chitin	95
Glomus mosseae	*Vicia faba*	6	Microscopy	60
Glomus mosseae	*Allium porrum*	10	Estimate	65
Glomus fasciculatum	*Glycine max*	16	Chitin	70
Glumus fasciculatum	*Sorghum bicolor*	3	Chitin	50
Ectomycorrhizae	*Abies amabilis*			
	23 years	12	Sampling plus estimate	
	180 years	3	for mantle percentage	
	Pseudotsuga menziesii	15	Microscopy plus harvest	

trees. Table 11.4 shows that 4–14% of the total carbon photosynthesized by the plant was allocated to the symbiont in a number of carbon-14 studies. Non-carbon-14 studies, using root harvest and washing at different times of the year, have shown that up to 50% of the turnover of carbon in a forest community could be attributed to the turnover of the fine roots and their fungal component. Such high values, however, have not been confirmed with carbon-14 labeling.

The explanation for the balance between increased growth attributable to better nutrition and the needs of the microbial partners can best be determined with carbon-14 distribution experiments. The distribution of carbon to above- and belowground parts of a soybean–*Glomus* association is shown in Fig. 11.9. Approximately 50% of the net photosynthate in these plants was found in the shoots, 6% was released during night respiration by the shoots, and 42% was translocated underground. This was nearly evenly distributed (12, 13 and 17%) to the nodules, the root, and the mycorrhizal fungi. Root growth accounted for 8% of the plant carbon; root respiration accounted for another 5%. The nodules immobilized 2% and respired 10% of the fixed carbon-14. In VAM, the fungi are composed of intraradical hyphae within the root, and extraradical hyphae in the soil. Figure 11.9 shows equal weights of the two components, representing 2.8% of the plant carbon-14 and accounting for 14% of the carbon-14 respiration.

The high microbial respiration losses, relative to low biomass production, differentiate rhizosphere organisms from the rest of the soil microflora (Harris and Paul, 1987). The normal soil flora have a high efficiency of incorporation of substrate carbon into their own biomass (see Chapters 5 and 6). However, symbiotic microorganisms could not carry out their symbiotic functions if the carbon primarily went to their own growth. The costs of N_2 fixation are high, and the apparent cost of phosphorus uptake and translocation by the associated mycorrhizal fungi also require that the majority of the carbon received by the microbial symbiont be spent on nutrient uptake rather than growth.

A number of studies have shown that plants can increase their photosynthetic rate to compensate for the needs of the microbial partners. This can come about both through increased leaf surface area in young plants and an increase in the amount of CO_2 fixed per unit weight of leaf. Studies of the ability of different plant associations to compensate for the needs of the mycorrhizal partners and of the efficiency of the microbial symbionts should continue to prove a useful area of research. A great variability in types of fungi and rhizobia has been determined. This natural heterogeneity can be further increased by molecular genetics, indicating the possibilities of enhancing the nutrient uptake through biological means.

Table 11.4

Effects of Vesicular–Arbuscular Mycorrhiza on Allocation and Photosynthesis

Fungus	Host	Carbon allocation to symbiont (%)	Carbon uptake[a] (increased photosynthesis) (%)
Glomus mosseae	*Vicia faba*	10	21
Glomus mosseae	*Vicia faba*	5	8
Glomus fasciculatum	*Glycine max*	14	20
Glomus fasciculatum	*Sorghum bicolor*	4	−21
Glomus mosseae	*Allium porrum*	12	13

[a]Photosynthetic rate as carbon uptake per gram leaf dry weight.

Figure 11.9. Distribution of photosynthate in a 6 week old soybean–*Glomus* association (net uptake, 131 mg carbon-14 day^{-1}). The sizes of the various components are shown within the boxes as percent of net uptake. The fluxes are shown on the arrows. ER VAM, Estimates of external root hyphae; IR VAM, intra root hypha.

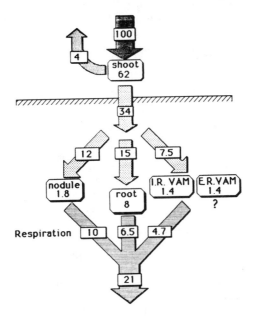

Methods of Study

The extent of VAM fungal infection is usually measured by observation of the percentage of the individual root fragments infected. This involves microscopic observation after clearing the roots with KOH and staining with dyes such as fuchsin red. The results, however, are difficult to interpret in terms of fungal carbon. In a system containing mycorrhizal fungi and plants in sand or vermiculite, it is possible to measure the amount of fungal tissue by determining the glucosamine content. Glucosamine, a component of chitin, is present in fungal cell walls but not in those of plants. The presence in soil of nonmycorrhizal fungi, which also contain glucosamine, however, makes it impossible to utilize this technique in the field unless $^{14}CO_2$ is used to label the VAM glucosamine in short-term exposures.

Carbon-14 translocation to the hyphae has usually been measured by autoradiography or hand-picking of hyphal bits followed by scintillation counting. Phosphorus uptake is measured by adding a phosphorus-32 solution to the soil. Counting of the radioactivity makes it possible to calculate the uptake of solution phosphorus relative to the uptake of other phosphorus forms, derived from organic or insoluble phosphorus. Split-root techniques, in which the fungal hyphae are placed in a phosphorus-32 solution, measures translocation to the plant.

Spores of VAM fungi range in size from 10 to 400 μm. They are separated from soil by flotation and aeration techniques in 50% glycerol, which has a specific gravity of 1.2. Alternatively, density gradient centrifugation can be utilized to separate the spores from the mycelium and organic debris. VAM fungi have not been grown in pure culture. Therefore, growth can only be studied in conjunction with a living plant host. This greatly hinders the biochemical and molecular biology investigations of these fungi. It also places limitations on the possible production of inoculum.

References

Allen, M. F., Moore, T. S., and Christensen, M. (1980). Phytohormone changes in *Bouteloua gracilis* infected by vesicular arbuscular mycorrhizae. I. Cytokinin increases in host plant. *Can. J. Bot.* **58**, 371–374.

Brown, M. F., and King, E. J. (1982). Morphology and histology of vesicular arbuscular mycorrhizae. *In* "Methods and Principles of Mycorrhizal Research" (N. C. Schenk, ed.), pp. 15–22. Am. Phytopathol. Soc., St. Paul, Minnesota.

Harley, J. L., and Smith, S. E. (1983). "Mycorrhizal Symbiosis." Academic Press, London.

Harris, D., and Paul, E. A. (1987). Carbon requirements of vesicular–arbuscular mycorrhizae. *In* "Ecophysiology of VA Mycorrhizal Plants" (G. R. Safir, ed.). CRC Press, Boca Raton, Florida.

Hayman, D. S. (1987). VA mycorrhizas in field crop systems. *In* "Ecophysiology of VA Mycorrhizal Plants" (G. R. Safir, ed.), pp. 171–192. CRC Press, Boca Raton, Florida.

Kinden, D. A., and Brown, M. F. (1976). Electron microscopy of vesicular arbuscular mycorrhizae of yellow poplar. IV. Host endophyte interactions during arbuscular deterioration. *Can. J. Microbiol.* **22**, 64–75.

Menge, J. A., Labanauskas, C. K., Johnson, E. L. V., and Platt, R. G. (1978). Partial substitution of mycorrhizal fungi for P fertilization in the greenhouse culture of citrus. *Soil Sci. Soc. Am. J.* **42**, 926–930.

Molina, R., and Trappe, J. M. (1982). Applied aspects of ECM. *In* "Advances in Agricultural Microbiology" (N. S. Subba Rao, ed.), pp. 305–324. Butterworth, London.

Pacovsky, R. S., Bethlenfalvay, G. J., and Paul, E. A. (1986). Comparisons between P-fertilized and mycorrhizal plants. *Crop Sci.* **26**, 151–156.

Sanders, F. E., Mosse, B., and Tinker, P. B. (1976). "Endomycorrhizas." Academic Press, London.

Trappe, J. M. (1987). Phylogenetic and ecological aspects of mycotrophy in the angiosperms from an evolutionary standpoint. *In* "Ecophysiology of VA Mycorrhizal Plants" (G. R. Safir, ed.), pp. 5–26. CRC Press, Boca Raton, Florida.

Trappe, J. M., and Berch, S. M. (1985). The prehistory of mycorrhizae. *In* "Mycorrhizae" (R. Molina, ed.), pp. 2–11. For. Res. Lab., Corvallis, Oregon.

Trappe, J. M., and Schenk, N. C. (1982). Taxonomy of the fungi forming endomycorrhizae. *In* "Methods and Principles of Mycorrhizal Research" (N. C. Schenk, ed.), pp. 1–9 Am. Phytopathol. Soc., St. Paul, Minnesota.

Supplemental Reading

Barea, J. M., and Azzon-Aguilar, C. (1983). Mycorrhizas and their significance in nodulating nitrogen-fixing plants. *Adv. Agron.* **36**, 1–54.

Brownlee, C., Duddridge, J. A., Malibari, A., and Read, D. J. (1983). The structure and function of mycelial systems of ECM roots with special reference to their role in forming interplant connection and providing pathways for assimilate and water transport. *Plant Soil* **71**, 433–443.

Fogel, R., and Hunt, G. (1979). Fungal and arboreal biomass in a western Oregon Douglas fir ecosystem: Distribution patterns and turnover. *Can. J. For. Res.* **9**, 245–256.

France, R. C., and Reid, C. P. P. (1983). Interactions of N and C in the physiology of ECM. *Can. J. Bot.* **61**, 964–984.

Hall, J. R., and Abbott, L. K. (1981). "Photographic Slide Collection Illustrating Features of the Endoganaceae," 3rd ed., pp. 1–27. Imermay Agricultural Research Centre and Soil Science Department, University of Western Australia, Perth.

Molina, R., ed. (1985). "Mycorrhizae." For. Res. Lab., Corvallis, Oregon.

Mosse, B., Stribley, D. P., and LeTacon, F. (1981). Ecology of mycorrhizae and mycorrhizal fungi. *Adv. Microb. Ecol.* **5**, 147–210.

Powell, C. L. (1982). Mycorrhizae. *In* "Experimental Microbial Ecology" (R. J. Burns and J. H. Slater, eds.), pp. 447–471. Blackwell, Oxford.

Reid, C. P. P. (1984). Mycorrhizae: A root/soil interface in plant nutrition. *ASA Spec. Publ.* **47**, 29–50.

Rothwell, F. M. (1984). Aggregation of surface mine soil by interaction between VAM fungi and lignin degradation products of lespedeza. *Plant Soil* **80**, 99–104.

Ruehle, J. L., and Marx, D. H. (1979). Fiber, food, fuel and fungal symbionts. *Science* **206**, 419.

Tinker, P. B. H. (1982). Mycorrhizas: The present position. *Proc. Int. Cong. Soil Sci. 12th, 1982*, pp. 150–166.

Chapter 12

Phosphorus
Transformations in Soil

Introduction

Phosphorus does not show biologically induced fluxes to and from the
atmosphere, as do carbon and nitrogen, nor, in contrast to reduced forms
of carbon and nitrogen, does it serve as a primary energy source for microbial oxidations. Nevertheless, soil organisms are intimately involved
in the cycling of soil phosphorus. They participate in the solubilization of
inorganic phosphorus and in the mineralization of organic phosphorus.
They are important in the immobilization of available soil phosphorus.
Although the microbial biomass is less than that of the higher plants, its
phosphorus content (in percent) may be as much as 10 times higher. Also,
within a single year there are several to many generations of growth by
diverse groups of soil organisms. Consequently, their annual uptake of
phosphorus often exceeds that of the higher plants. Fortunately, microbial
immobilization of phosphorus is not of long duration, and on balance it
may be beneficial to plants. Short-term biological immobilization spares
some soil phosphorus from long-term fixation in minerals.

Global Aspects of the Phosphorus Cycle

Phosphorus exists in nature in a variety of organic and inorganic forms
but primarily in either insoluble or only very poorly soluble inorganic
forms. The total amount of phosphorus in the earth's crust is of the order
of 10^{15} metric tons. It exists mainly as apatites, with the basic formula
$M_{10}(PO_4)_6X_2$. Commonly the mineral (M) is calcium, and the anion (X) is
fluorine. The anion can also be Cl^-, OH^-, or CO_3^{2-}; thus there exists

fluor-, chloro-, hydroxy-, and carbonate apatites. Diverse substitutions and combinations of M and X result in some 200 forms of phosphorus occurring in nature. Rock phosphates high in carbonate apatite are most commonly mined as fertilizer sources.

Estimates of the amounts of phosphorus lodged in certain terrestrial and oceanic reservoirs are shown in Table 12.1. By far the largest reservoir is oceanic sediment; this constitutes a sink and a small but steady drain on terrestrial phosphorus released by weathering and/or the biota. Erosion and the flushing of human wastes to the ocean are the major sources for the movement of terrestrial phosphorus to the oceanic. Plant phosphorus is largely recycled by microbial mineralization of litter and other organic debris. Long-term cultivation without supplemental phosphorus leads to depletion of soil phosphorus. Data compiled by Tiessen et al. (1982) exemplify such loss (Table 12.2). Continuing additions of either fertilizer phosphorus or of animal manures cause enrichment of soil phosphorus; they also cause increase in soil carbon and nitrogen.

There is good evidence that phosphorus is the dominant element controlling carbon and nitrogen immobilization. In a classical and much discussed paper, Redfield (1958) hypothesized that phosphorus controls the carbon, nitrogen, and sulfur cycles of marine systems. He noted that the oceanic carbon:nitrogen:phosphorus (C:N:P) ratio paralleled that of the plankton and believed the following general relationships to occur:

$$106 \; CO_2 + 16 \; NO_3^- + PO_4^{3-} + 122 \; H_2O + 18 \; H^+ \xrightarrow[\text{trace elements}]{\text{Solar energy}}$$

$$C_{106}H_{263}O_{110}N_{16}P_1 + 138 \; O_2$$

Table 12.1
Major Reservoirs of Phosphorus in the Earth[a]

Reservoir	Total phosphorus ($\times$ 10^{12} kg)
Land	
Soil	96–160
Mineable rock	19
Biota	2.6
Fresh water (dissolved)	0.090
Ocean	
Sediments	840,000
Dissolved (inorganic)	80
Detritus (particulates)	0.65
Biota	0.050–0.12

[a]From Bolin et al. (1983).

Table 12.2

Changes in Phosphorus Content by Long-Time Cultivation of soils from Three Grassland Soil Associations of the Canadian Prairies[a]

Soil association	Native prairie	Cultivated soil[b]	Loss (%)
Blain Lake			
Organic carbon (mg g^{-1})	47.9 ± 10.2	32.8 ± 5.2	32
Total phosphorus (μg g^{-1})	823 ± 92	724 ± 53	12
Organic phosphorus (μg g^{-1})	645 ± 125	528 ± 54	18
Inorganic phosphorus (μg g^{-1})	178 ± 47	196 ± 8	Not significant
Southerland			
Organic Carbon (mg g^{-1})	37.7 ± 6.5	23.7 ± 1.8	37
Total phosphorus (μg g^{-1})	756 ± 28	661 ± 31	12
Organic phosphorus (μg g^{-1})	492 ± 5.2	407 ± 30	17
Inorganic phosphorus (μg g^{-1})	256 ± 44	254 ± 19	Not significant
Bradwell			
Organic carbon (mg g^{-1})	32.2 ± 8	17.4 ± 1.6	46
Total phosphorus (μg g^{-1})	746 ± 101	527 ± 15	29
Organic phosphorus (μg g^{-1})	446 ± 46	315 ± 21	29
Inorganic phosphorus (μg g^{-1})	300 ± 84	212 ± 46	29

[a]Adapted from Tiessen *et al.* (1982).
[b]Cultivation was for 90, 75, and 60 years, respectively.

The C:N:P ratios for a number of terrestrial situations are shown in Table 12.3. The aquatic algae and soil bacteria have a similar C:N ratio of roughly 6:1. The algae have lower C:P ratios than the bacteria, but fall within the range for C:P found in soil organic matter (SOM).

The early work of Walker (1965) and coworkers led to the hypothesis that accumulation of carbon, nitrogen, sulfur, and phosphorus in SOM is dependent on the phosphorus content of the soil parent material. This has been verified by farming practices in Australia and New Zealand in which fertilizer phosphorus application has increased the fertility of managed grasslands and increased total organic matter content. The available phosphorus content of many heavily fertilized cultivated soils such as the maize–soybean rotations in parts of the USA–Midwest is now higher than required for most crops. To date nobody has indicated that the higher phosphorus is resulting in higher organic carbon and nitrogen accumulations as was found for the grasslands of New Zealand. These soils could prove to be a good test case for the application of the Redfield–Walker hypotheses to cultivated soils.

Table 12.3
Carbon: Nitrogen: Phosphorus (C:N:P) Ratios in
Organisms and Soil Organic Matter

Natural component	C	N	P
Marine algae	106	16	1
Soil bacteria	31	5	1
Grassland soil			
Virgin	191	6	1
Cultivated, fertilized	119	9	1
Weakly weathered soil	80	5	1
Strongly weathered soil	200	10	

Inorganic Forms

Inorganic forms of phosphorus occur in combination with iron, aluminum, calcium, fluorine, or other elements. The hundreds of combinations of phosphorus in soil necessitate discussion of them collectively, and on the basis of extractability rather than individually and structurally. As stated earlier, most soil phosphates are insoluble or very poorly soluble. When phosphorus is added to soil as a soluble salt, it becomes fixed or bound to the extent that very little of the added phosphorus is reextractable with water. Nor is the major fraction of the retained phosphorus extractable by dilute acids or bicarbonate solution. This portion of the retained phosphorus is commonly designated as fixed phosphorus. The portion that is extractable by dilute acid or bicarbonate is designated as available phosphorus; it is considered to be the amount of soil phosphorus available for uptake by living organisms. Available phosphorus thus measured is partly inorganic and partly organic. That portion of the total phosphorus in soil that is resin extractable is designated as labile phosphorus, defined as that fraction of the soil phosphorus that can enter solution by isoionic exchange during a given time span.

Solution Phosphorus

The concentration of phosphorus in the soil solution is of the order of 0.1 to 1 ppm; only rarely does it exceed this range. Phosphorus solubility is complicated by common ion–ion association and pH effects, and by the amount of phosphorus adsorbed on the surfaces of clay minerals. Solution phosphorus equilibrates rapidly with the labile phosphorus, and for most soils the ratio of labile inorganic phosphorus to solution phosphorus is linear.

Plants meet their phosphorus requirements from the solution pool. Metabolic energy is expended to move phosphorus into the cytoplasm, where the phosphorus concentration may be 50–100 times higher than in the soil solution. Calculation of phosphorus uptake in the field indicates that uptake rates of 1 and 2 μmol phosphorus per gram of root per day will maintain adequate phosphorus concentrations in plants growing at rates of 8 to 10% per day. A minimum level of phosphorus must be attained in the root before plant translocation can occur. No translocation of phosphorus occurs in grasses below a minimum level of approximately 0.05% and translocation is proportional to root phosphorus concentration, between the minimum level and an upper limit of approximately 0.5%; further increases in root phosphorus concentration do not further increase translocation.

Root uptake from the solution pool may involve renewal of this pool as often as 50 to 250 times per day during periods of plant growth under nearly ideal conditions. Soil organisms contribute to the renewal of the solution pool both by their solubilizing effects on the labile phosphorus and by their mineralization of organic phosphorus.

Solubilization of Inorganic Phosphorus

Soil organisms and plant roots participate in the solubilization of soil phosphorus, mainly through their production of CO_2 and organic acids. Measurements of the precise amount of soil phosphorus solubilized by either roots or soil organisms is complicated by the concomitant mineralization of organic phosphorus. Prior to the widespread commercial production of superphosphates by acidification with mineral acid, the microbial solubilization of phosphate was enhanced by mixing manure with ground phosphate rock and allowing an incubation period prior to field application. Attempts were made to isolate the specific bacteria responsible for solubilization and to use such organisms as soil or seed inoculants. Most frequently isolated were species of *Pseudomonas* and *Bacillus*. In past years, *B. megaterium* var. *phosphaticum* was widely used in a bacterial inoculant known as phosphobacterin. Although heavily used in the Soviet Union over several decades, good evidence of its effectiveness was never achieved. Lack of benefit from an added inoculant does not imply that soil organisms, taken collectively, are unimportant as solubilizing agents.

Organic Forms

The organically bound phosphorus usually constitutes 30–50% of the total phosphorus in most soils, although it may range from as low as 5% to as high as 95%. Table 12.4 illustrates the total and organic phosphorus content in eight grassland soils in the western United States. Organic phosphorus occurs in soils principally as phytates, nucleic acids and their derivatives, and phospholipids. Phytin is largely inositol hexaphosphate (Fig. 12.1); it is synthesized both by microorganisms and plants and is the most stable of the organic forms of phosphorus in soil. Phytin was found to account for 17% of the total organic phosphorus in 49 Iowa soils, with individual values ranging from 3 to 52%. Half or more of the total organic phosphorus in soil has not been structurally identified. Inositol ring compounds carrying from one to five atoms of phosphorus also occur in soil; these possibly represent degradation products of inositol hexaphosphate.

Evidence concerning the occurrence of nucleic acid phosphorus in SOM is largely indirect. Source materials are contributed by soil organisms and plant litter; constituent parts of nucleic acid molecules are identifiable in hydrolysates of soil extracts. Among identifiable components are cytosine, adenine, guanine, uracil, hypoxanthine, and xanthine; the two last named are decomposition products of guanine and adenine. Of the total organic phosphorus in soil, only about 1% can be identified as nucleic acids or their derivatives. The susceptibility of nucleic acids to decomposition, together with a lack of incorporation into stable organic matter, is believed to be responsible for their low level of persistence in soil.

Table 12.4
Total and Organic Phosphorus in the Surface 10 cm of
Soil at North American Grassland Sites

Site	Phosphorus ($\mu g\ g^{-1}$ soil)	
	Total	Organic
Cottonwood, South Dakota	554	310
Bridger, Montana	1234	675
Osage, Kansas	251	227
Ale, Washington	748	29
Jornado, New Mexico	445	37
Pantex, Texas	94	239
Bison, Montana	835	582
Pawnee, Colorado	345	131

Figure 12.1. Organic phosphorus compounds. (A) Inositol. If given six phosphorus substitutions on the ring (C—P linkages), it becomes inositol hexaphosphate. Fewer substitutions yield 1-, 2-, 3-, 4-, or 5-phosphatidyl inositol phosphates. (B) Phosphoglycerides (C—O—P linkages). One is shown linked to serine. (C) Phosphate sugars. One is shown linked to uridine. (D) Nucleic acid components.

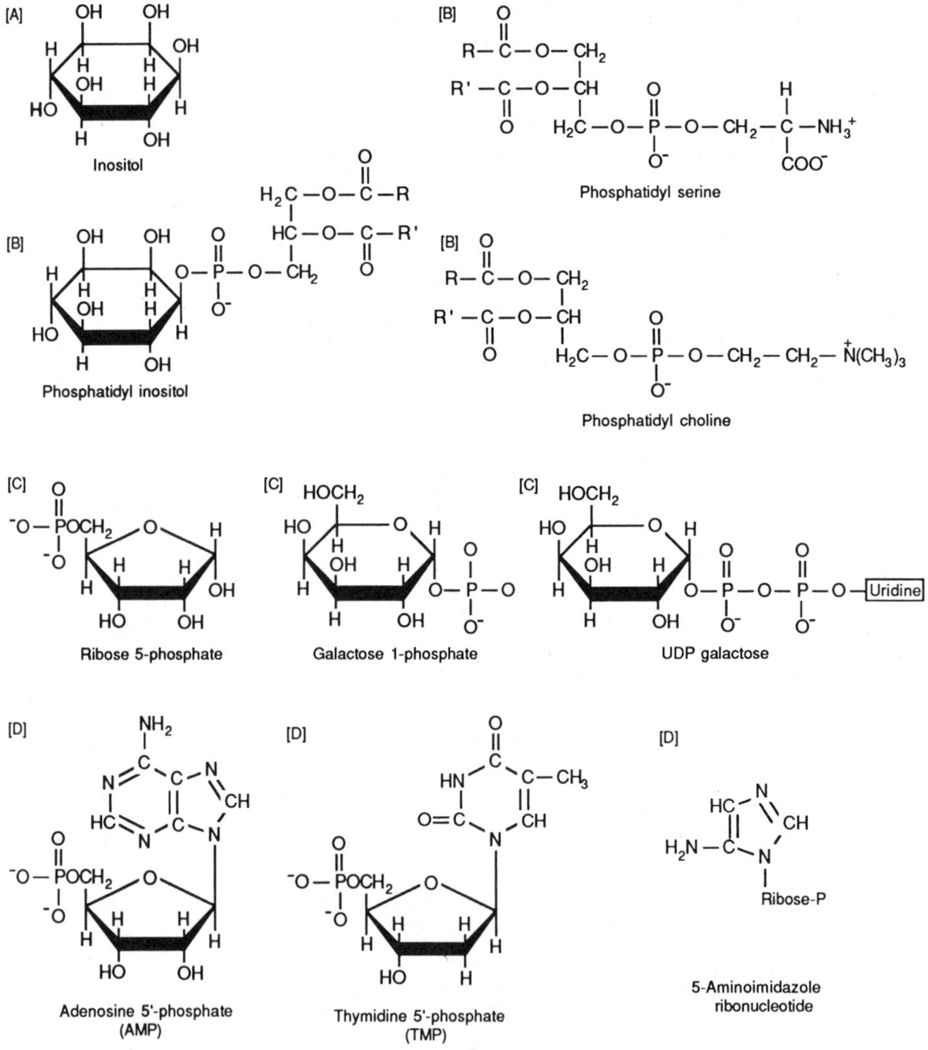

Organic phosphorus occurring in alcohol and ether extracts of soil is indicative of the presence of phospholipids. Choline has been identified; it is one of the products of hydrolysis of lecithin. Various studies have shown that only about 1 to 5 ppm of phospholipid phosphorus occurs in soil, although values as high as 34 ppm have been encountered. In a study on chernozemic soils in which two-thirds or more of the total organic phosphorus remained unidentified, 1–2% was reported as phospholipid phosphorus. Even smaller amounts of sugar phosphates are found in soil. These are also easily decomposable, and therefore, together with the nucleic acids and phospholipids, cannot be considered as contributing significantly to the estimates of 350 to 2000 years as the mean residence time for total organic phosphorus in soil.

The phosphorus within the soil biomass can be determined after lysis of the cells with $CHCl_3$. Extraction of this phosphorus with $NaHCO_3$, and relating its value to the phosphorus extracted from an aliquot not treated with $CHCl_3$, yields an estimate of microbial phosphorus. Work with soils having a pH of 6.2 to 8.2 showed a recovery K_p of 40% of added ^{32}P-labeled cells, suggesting a calculation of microbial phosphorus as follows:

$$\text{Microbial P} = \frac{CHCl_3 \text{ P released}}{K_p} = \frac{CHCl_3P}{0.4}$$

Fluxes of Phosphorus in the Plant Root–Soil System

The very low level of phosphorus in soil solution and the necessity for its frequent renewal suggests that the transfer of SOM phosphorus to inorganic phosphorus is primarily, but not exclusively, microbially mediated. Plant roots also produce phosphatases, and plants are known to differ in the amounts that they produce. Such differences may be one of the determinants in plant succession. Also, some plant species are known to produce more extracellular phosphatases in response to low levels of available phosphorus.

Microbial phosphatases are produced by most organotrophic members of the soil microflora. Microbial mineralization of organic phosphorus is strongly influenced by environmental parameters. Mild alkalinity favors mineralization, and calcareous soils can be expected to have lower organic phosphorus content than acid soils. Temperature is also influential; thus some pastures respond to phosphorus fertilization in the early spring but not later, when the soil temperature permits phosphorus mineralization at a rate adequate to meet the plant requirement.

A simulation of the flux of organic phosphorus in a grassland system

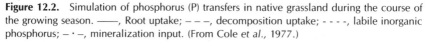

Figure 12.2. Simulation of phosphorus (P) transfers in native grassland during the course of the growing season. ——, Root uptake; – – –, decomposition uptake; - - - -, labile inorganic phosphorus; – · –, mineralization input. (From Cole *et al.*, 1977.)

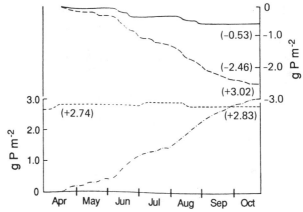

during a single growing season, and based on the best available data for the site, is shown in Fig. 12.2. This is based on the concept of interactions of the various forms of soil phosphorus shown in Fig. 12.3. More of the mineralized phosphorus is used by the soil organisms than by the plants. This is understandable; the microflora has a phosphorus content of 1.0 to 2.0%, and the plants, 0.05 to 0.10%. The constancy in the amount of labile inorganic phosphorus during the season emphasizes the importance of the mineralization process in meeting the phosphorus requirement in plants. Transfers and transformations are most important in crop production and ecosystem functioning.

One microbial form, the mycorrhyzal fungi, plays a more important role in phosphorus transfers than the general soil population. Three mechanisms are probably involved. First, the mycorrhizal hyphae contribute to the solubilization of mineral phosphate by their production of respiratory CO_2 and by excretion of organic acids. Thus there is a mycorrhizosphere effect that supplements the rhizosphere effect. Second, the mycorrhizal hyphae extend soil exploration over and beyond the soil exploration accomplished by the plant roots themselves. Fungal hyphae penetrate SOM particles and macroaggregates more thoroughly than do root extensions. Third, the hyphae may take up phosphorus at lower levels of concentration in the soil solution than can the roots. This capability can be especially critical for seedlings developed from seeds with very small food reserves, and thus with limited capability for soil exploration by newly developing roots.

Figure 12.3. Compartments and flows of phosphorus. (From Stewart and McKercher, 1982.)

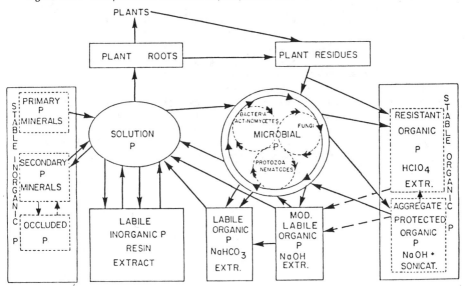

References

Bolin, B., Rosswell, T., Freney, J. R., Ivanov, M. V., and Richey, J. E. (1983). C, N, P, and S cycles, major reservoirs and fluxes. *In* "The Major Biogeochemical Cycles and Their Interactions" (B. Bolin and R. B. Cook, eds.). Wiley, and sons, New York.

Cole, C. V., Innis, G. S., and Stewart, J. W. B. (1977). Simulation of P cycling in semi-arid grasslands. *Ecology* **58**, 1–15.

Redfield, A. C. (1958). Biological control of chemical factors in the environment. *Am. Sci.* **46**, 205–221.

Stewart, J. W. B., and McKercher, R. B. (1982). Phosphorus cycle. *In* "Experimental Microbial Ecology" (R. G. Burns and J. H. Slater, eds.), pp. 221–238. Blackwell, Oxford.

Tiessen, H., Stewart, J. W. B., and Bettany, J. R. (1982). Cultivation effects on the amounts and concentration of carbon, nitrogen and phosphorus in grassland soils. *Agron. J.* **74**, 831–835.

Walker, T. W. (1965). The significance of phosphorus in pedogenesis. *In* "Experimental Pedeology" (E. A. Hallsworth and D. V. Crawford, eds.), pp. 295–315. Butterworth, London.

Supplemental Reading

Borie, F., and Zuniono, H. (1983). Organic matter P association as a sink in P fixation processes in allophanic soils of Chile. *Soil Biol. Biochem.* **15**, 599–603.

Coleman, D. C., Reid, C. P. P., and Cole, C. V. (1983). Biological strategies of nutrient cycling in soil systems. *Adv. Ecol. Res.* **13**, 1–55.

Gray, L. E., and Gerdemann, J. W. (1969). Uptake of phosphorus-32 by vesicular–arbuscular mycorrhizae. *Plant Soil* **30**, 415–422.

Hedley, M. S., and Stewart, J. W. B. (1982). Method to measure microbial phosphate in soils. *Soil Biol. Biochem.* **14**, 377–385.

Larsen, S. (1967). Soil phosphorus. *Adv. Agron.* **19**, 151–210.

Olson, S. R., and Watanabe, F. S. (1963). Diffusion of soil P as related to soil texture and plant uptake. *Soil Sci. Soc. Am.* **29**, 154–158.

Stevenson, F. J. (1986). "Cycles of Soil." Wiley, New York.

Stewart, T. J. (1984). Interrelation of C, N, sulfur and P cycles during decomposition processes in soil. *In* "Current Perspectives in Microbial Ecology" (M. J. Klug and C. A. Reddy, eds.), pp. 442–446. Am. Soc. Microbiol., Washington, D.C.

Thompson, L. M., Black, C. A., and Zoellner, J. A. (1954). Occurrence and mineralization of organic P in soils with particular reference to associations with N, C and pH. *Soil Sci.* **77**, 185–196.

Chapter 13

Sulfur Transformations in Soil

Introduction

Recognition of the existence in soil of lithotrophic bacteria capable of oxidizing sulfur occurred almost simultaneously with the discovery of the nitrifying bacteria. Winogradsky (1887, 1889) recognized that certain bacteria were able to obtain their growth energy by oxidizing sulfur to SO_4^{2-}. Major emphasis was given in following decades to the comparative physiology of the organisms, with little attention to their role in the sulfur cycle in soil. There have been a number of studies of sulfur cycling, with many of them linked to studies of nitrogen cycling. Sulfur, together with nitrogen, is an integral part of proteins. On the average, in proteins, there are 36 atoms of nitrogen for each atom of sulfur. Both are absorbed by plants as water-soluble anions, and both have major reserves of potentially mineralizable compounds in soil organic matter (SOM). Also, both are subject to atmospheric deposition, leaching to ground water, and to microbial reduction of a soluble available form of the nutrient to a gaseous form that is lost to the atmosphere. Not surprisingly, many of the conceptual cycles and transfers for sulfur are similar to those for nitrogen (Fig. 13.1).

Most of the early studies of the microbial transformations of sulfur in nature were concerned with the geochemistry of marine sediments and the formation of sulfide ores. Problems with sulfur deficiencies and excesses in soils only slowly gained attention. Additions of nitrogen and potassium as sulfate salts and of phosphorus as single superphosphate $[Ca(H_2PO_4)_2CaSO_4]$ often met an unrecognized crop requirement for supplemental sulfur. Increases in use of high-analysis fertilizers, low or lacking in sulfur, have led to recognition that sulfur-deficient arable soils exist.

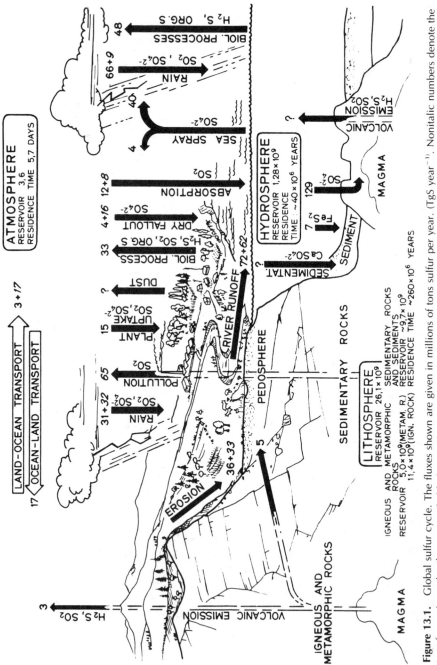

Figure 13.1. Global sulfur cycle. The fluxes shown are given in millions of tons sulfur per year. (TgS year⁻¹.) Nonitalic numbers denote the transfers estimated to have prevailed before civilization had a significant influence on the sulfur cycle; *italic* numbers give the amounts that humans have added by various activities. (From Zehnder and Zinder, 1980.)

In North America, sulfur-deficient soils occur in parts of the Pacific Northwest and in some wooded soils in Canada.

Sulfur in excess concentration in soil is harmful to plants. Acid rain, containing SO_4^{2-} and NO_3^-, has great potential for acidifying lakes and for damaging forests. Water in equilibrium with atmospheric CO_2 has a pH of 5.7 at 25°C, but only 1.5 ppm of SO_2 sulfur is required to lower the pH of rainfall to 4.0. Some 65 million metric tons of pollution sulfur enter the atmosphere on an annual basis (Fig. 13.1). This sulfur, however, is concentrated in the air downwind from heavily industrialized areas and is found in the atmosphere as SO_2, SO_4^{2-}, aerosols, and airborne particulates. Areas with no pollution receive approximately 1 kg ha^{-1} year^{-1}. Areas with some pollution, such as Iowa or North Carolina, receive 10 kg ha^{-1} year^{-1}. Polluted areas near cities can receive 100 kg ha^{-1} year^{-1}; 15–20% of this can come down as H_2SO_4. The remainder is sorbed as SO_2 or received as particulates. Many soils near cities absorb more SO_2 from dry than from wet deposition.

The dissolution of carbonates, aluminum silicates, and oxides of iron, aluminum, and manganese occurs slowly, and many years may elapse before major plant injuries are noted. In forests, the slow disappearance of lichenous colonizations on tree bark usually precedes the onset of foliar damage. Excess sulfur in anaerobic sandy soils in which available cations have been depleted by cropping may lead to sulfide remaining in solution. This in turn accentuates physiological disorders, such as the *akiochi* disease of rice. One corrective procedure involves substitution of Cl^- for SO_4^{2-} as the carrier for NH_4^+ or K^+ in fertilizer. In localized areas, acid rain and sorbed SO_2 may intensify soil weathering processes. Acid rain causes the corrosion of building materials; the most publicized is that done to some of the great European cathedrals and to the Grecian marbles. Acidification of lakes in North America and the destruction of extensive forests in Europe are linked to acid rain. The forms and reactions of sulfur in nature are thus the object of intense scrutiny.

Forms of Sulfur in Nature

Table 13.1 shows the distribution of sulfur in some representative world soils. Most agricultural soils contain sulfur in the range 20–2000 μg g^{-1}. Many volcanic ash, organic, and tidal-marsh soils contain 3000 or more μg g^{-1}, and some desert soils have a total sulfur content in excess of 10,000 μg g^{-1}. Sulfur occurs in soil in organic and inorganic forms. In soils other than aridisols, the inorganic fraction is typically small compared to the organic.

Table 13.1

Amounts and Distribution of Sulfur in Some World Soils

| | | | Total sulfur (metric tons) | | |
| | | | | Organic | |
Location	Type of soil	Total sulfur ($\mu g\ g^{-1}$)	Inorganic, reducible	Carbon bonded	Ester sulfate
Saskatchewan	Agricultural	88–760	0.5–13	29–59	41–71
British Columbia	Grassland	286–928	ND	31–61	39–69
	Forest	162–2328	ND	20–47	53–80
	Organic	1122–30,430	ND	28–75	25–72
	Agricultural	214–438	2	18–45	55–82
Iowa	Agricultural	57–618	2–8	43–60	7–18
Carolinas	Tidal marsh	3000–35,000	—	—	—
Hawaii	Volcanic ash	180–2200	6–50	50–94	50–94
Eastern Australia	Agricultural	38–545	4–13	10–70	24–76
Nigeria	Agricultural	25–177	4–20	80–96	80–96
Brazil	Agricultural	43–398	5–23	20–65	24–59

Table 13.2
Oxidation States of Various Inorganic Sulfur Compounds

Form	Formula	Oxidation state(s)
Sulfate	SO_4^{2-}	+6
Sulfite	SO_3^{2-}	+4
Thiosulfate	$S_2O_3^{2-}$ ($^-S-SO_3^-$)	−2, +6
Tetrathionate	$S_4O_6^{2-}$	+2.5
Thiocyanide	$S-C-N^-$	−2
Trithionate	$^-O_3SSSO_3^-$	+6, +2, +6
Elemental	S^0	0
Disulfide	HS^-	−2
Sulfide	S^{2-}	−2

Inorganic Forms

Inorganic sulfur exists in nature in a number of oxidation states, ranging from +6 in SO_4^{2-} and its derivatives, to −2 in H_2S and its derivatives (Table 13.2). More than 2000 sulfur-bearing minerals with sulfur content ranging from 7 to 53% make sulfur the thirteenth most abundant element in the earth's crust. Weathering during soil formation occurs very slowly, and primary mineral sulfur is seldom a significant source of plant-available sulfur. Inorganic sulfur exists primarily as SO_4^{2-} in calcareous soils of semiarid and arid regions, in volcanic ash, and in recently reclaimed tidal-marsh areas. It may be present as gypsum ($CaSO_4$), as a basic aluminum sulfate, and as a contaminant of calcium carbonate in concentrations ranging as high as 3000 ppm.

Under tidal-marsh and saline-lake conditions, sulfur accumulates mainly as sulfides and polysulfides of iron. In acidic soils in areas of Southeast Asia, tropical Africa, and tropical South America, high SO_4^{2-} retention occurs as a consequence of precipitation and dissolution of aluminum hydroxy sulfates. Their iron analogs form compounds such as jarosite [$KFe_3(OH)_6(SO_4)_2$] and coquinbite [$Fe_2(SO_4)_3 \cdot 5H_2O$].

Organic Forms

Sulfur is an essential nutrient for all living systems. Animals other than ruminants can only incorporate sulfur when it is obtained in reduced form in amino acids. Plants and microorganisms can produce and utilize a variety of forms. In soil these forms find their way into SOM and can be classified into two major types. Carbon-bonded sulfur (Fig. 13.2) occurs in amino acids such as cysteine, cystine, and methionine, in cofactors such as biotin, thiamine, and coenzyme A, in iron–sulfur proteins (ferroxodins), and in

Figure 13.2.   Forms of carbon-bonded sulfur.

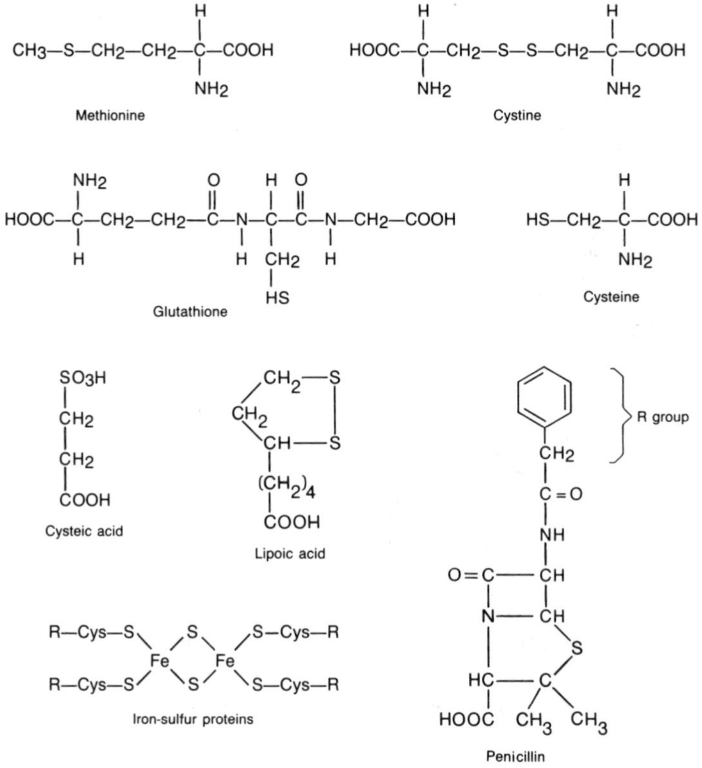

$$CH_3-S-CH_2-CH_2-\underset{\underset{NH_2}{|}}{\overset{\overset{H}{|}}{C}}-COOH$$

Methionine

$$HOOC-\underset{\underset{NH_2}{|}}{\overset{\overset{H}{|}}{C}}-CH_2-S-S-CH_2-\underset{\underset{NH_2}{|}}{\overset{\overset{H}{|}}{C}}-COOH$$

Cystine

$$HOOC-\underset{\underset{H}{|}}{\overset{\overset{NH_2}{|}}{C}}-CH_2-CH_2-\overset{\overset{O}{\|}}{C}-\underset{\underset{H}{|}}{\overset{\overset{H}{|}}{N}}-\underset{\underset{\underset{HS}{|}}{CH_2}}{\overset{\overset{H}{|}}{C}}-\overset{\overset{O}{\|}}{C}-\underset{\underset{H}{|}}{N}-CH_2-COOH$$

Glutathione

$$HS-CH_2-\underset{\underset{NH_2}{|}}{\overset{\overset{H}{|}}{C}}-COOH$$

Cysteine

$$\begin{array}{c} SO_3H \\ | \\ CH_2 \\ | \\ CH_2 \\ | \\ COOH \end{array}$$

Cysteic acid

Lipoic acid

Iron-sulfur proteins

$$\begin{array}{c} R-Cys-S \\ R-Cys-S \end{array} \diagdown \underset{Fe}{S} \diagup \overset{S}{\diagdown} \underset{Fe}{S} \diagdown \begin{array}{c} S-Cys-R \\ S-Cys-R \end{array}$$

R group

Penicillin

lipoic acids. The sulfur-containing amino acids maintain the secondary, tertiary, and quarternary structures of proteins via disulfide linkages. Many enzymes are inhibited when treated with reagents that destroy sulfhydryl groups. The sulhydryl groups are also involved in binding of substrates to enzymes; e.g., the sulfhydryl group of glyceraldehyde-3-phosphate dehydrogenase binds the substrate through a thioester bond.

In the second major type of soil organic sulfur, the sulfur is attached in the organic matrix via —O—, or sometimes, —N— bonds (Fig. 13.3). Collectively, they may be termed sulfate esters. The measurement of this form involves reduction with hydroiodic acid (HI), and they are often referred to as HI-reducible sulfur. This is considered to be the more biologically active or labile form in soil. It often is a storage product in plants and microorganisms when available sulfur is in excess. Glucosinolates can constitute a significant portion of the organic SO_4^{2-} in plants of the family Brassicaceae (alternate name, Cruciferae). Another important

Figure 13.3. Ester sulfate forms of organic sulfur found in soil and as constituents of organisms.

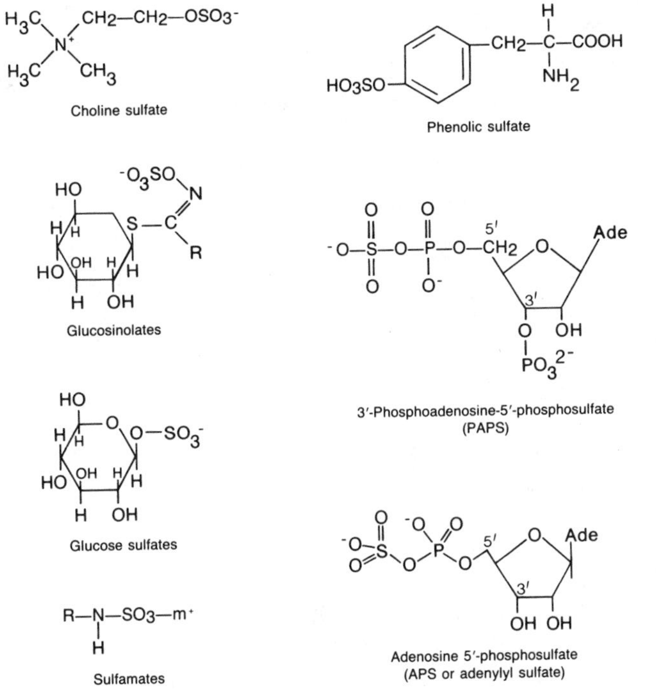

Choline sulfate

Phenolic sulfate

Glucosinolates

3'-Phosphoadenosine-5'-phosphosulfate
(PAPS)

Glucose sulfates

Sulfamates

Adenosine 5'-phosphosulfate
(APS or adenylyl sulfate)

group of sulfur compounds are the sulfatophosphates, such as adenosine 5-sulfatophosphate (APS) and adenosine 3-phosphate 5-sulfatophosphate (PAPS) (Fig. 13.3). These participate in many inorganic sulfur transformations.

Sulfur Constituents in Microorganisms

Biomass sulfur represents 2 to 3% of the total organic sulfur in soils. Microbial sulfur is measured by lysing the cells with chloroform ($CHCl_3$) and measuring the sulfur released to a $CaCl_2$ or $NaHCO_3$ extractant. Analysis of the sulfur content of soil microorganisms shows that the major fraction is present as amino acids. *Arthrobacter* and *Pseudomonas* contain roughly 10% of their sulfur in an oxidized form regardless of whether they are grown at low or high levels of substrate sulfur (Table 13.3). The fungi *Fusarium* and *Trichoderma* show a similar percentage when grown at low substrate sulfur levels, but roughly 40% is in the ester form when grown at high levels (16 $\mu g\ cm^{-3}$). These data are of particular interest because

Table 13.3
Total Sulfur and HI-Reducible Sulfur of Two Bacteria and Two Fungi Grown in Culture
with Varying Sulfur Concentration[a]

Organism	Total sulfur (ppm)			HI-reducible sulfur (%)		
	S_1^b	S_2	S_3	S_1	S_1	S_2
Arthrobacter globiformis	928	1123	1355	10	11	7
Pseudomonas cepacia	1108	1341	1339	13	10	12
Fusarium solani	1013	2318	2772	12	24	42
Trichoderma harzianum	852	1262	2034	6	17	37

[a]From Saggar *et al.* (1981).
[b]S_1, S_2, S_3 = 1, 4, and 16, μg of sulfur cm^{-3}, respectively, in the culture medium.

in the past it has not been possible to identify the source of ester sulfates
in SOM. Fungi have been reported to store intracellular sulfur as choline
sulfate, but otherwise little is known about the microbial storage of ester
sulfur compounds. Apparently, fungi retain most of the ester sulfur they
synthesize, whereas bacteria release a large proportion of the ester into
the culture medium. Sulfolipids, which represent another form of ester
sulfate, have been reported for a limited number of bacteria. These esters
are known to be produced in laboratory cultures but have not yet been
shown to be produced in soil. Algal sulfate esters also exist; the well-
known algal compound, agar, is a sulfuric acid ester of a linear galactan.

Other forms of sulfur, such as phenyl sulfates and elemental sulfur,
have been found in a variety of soil organisms. Elemental sulfur has been
identified in the sporocarps of ectomycorrhizal fungi as well as in other
self-inhibited and dormant structures. Elemental sulfur is deposited by
some of the bacteria capable of utilizing H_2S as a reducing agent during
photosynthesis.

Organic Sulfur Concentrations in Humus

The carbon:sulfur (C:S) ratio in SOM is not as consistent as is the car-
bon:nitrogen (C:N) ratio. Major differences are found due to type of parent
material, leaching, and sulfur inputs. Mollisols of the prairies have C:N:S
ratios of 90:8:1; luvisols and spodosols can range up to 200:12:1. Wide
ratios are usually found in areas of low sulfur supply. This results in low
net sulfur mineralization potentials. A general worldwide ratio could be
stated as 130:10:1.3, with wider ratios being found in native grass and
woodland soils. Excess sulfur as wet or dry pollutants can increase the

level of soil organic sulfur. The carbon-bonded sulfur is difficult to analyze; it is usually calculated by subtraction of the value for HI-reducible sulfur from total organic sulfur. Table 13.1 shows that HI-reducible sulfur accounts for 25 to 75% of the organic sulfur. The proportion of HI-reducible sulfur generally increases with profile depth; it is concentrated in the clay-sized fraction and appears in the fulvic acid when the soil is extracted with NaOH or $Na_2P_4O_7$.

Organic sulfur not reducible by HI is generally assumed to be carbon bonded. A subfraction of this portion can be reduced with Raney nickel. This is believed to represent the amino acid portion of the carbon-bonded sulfur.

Figure 13.4. Sulfur (S) transformations in nature. Elemental sulfur is shown as a storage product, and the possibility of SO_4^{2-} sorption in certain soils is included.

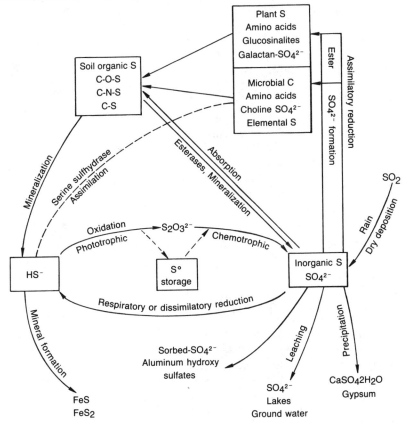

Transformations of Sulfur

Mineralization of Organic Sulfur

The transformation of C—O—S, C—N—S, and R—C—S compounds originating from plants or microorganisms can proceed through both aerobic and anaerobic pathways (Fig. 13.4). The hydrolysis of ester sulfates occurs by the splitting of the O—S bond by sulfatase enzymes:

$$R—O—SO_3^- + H_2O \xrightarrow{\text{sulfatase}} ROH + H^+ + SO_4^{2-}$$

There are numerous sulfatases, characterized by high specificity toward the organic part of the molecule and the sulfur moiety. At least three sulfatases, including two arylsulfatases and cholinesulfatase, are known to be produced by a pseudomonad. Unless SO_4^{2-} is released extracellularly when in excess of microbial growth requirements, the hydrolysis of ester sulfates microbially will not make sulfur available for plant growth. Studies lend support to the theory that several sulfatases are located outside the cytoplasm, but nevertheless, little correlation between sulfatase content with soil carbon has been found. Work on fractionation of SOM has indicated that total sulfur content, and especially the presence of ester sulfates, are not highly correlated with the potential for sulfur mineralization during incubation. Sulfur mineralization is less highly correlated with the degradation of carbon than is nitrogen mineralization.

The mineralization of amino acids such as cysteine can occur anaerobically via cysteine desulfhydrase or serine sulfhydrase:

$$\text{HSCH}_2—\text{CH}—\text{COOH} \xrightarrow[\text{cysteine desulfhydrase}]{\text{H}_2\text{O}} \text{H}_2\text{S} + \text{NH}_3 + \text{CH}_3\text{COOH}$$
$$|$$
$$\text{NH2}$$

$$\text{HSCH}_2—\text{CH}—\text{COOH} \xrightarrow[\text{serine sulfhydrase}]{\text{H}_2\text{O}} \text{HOCH}_2—\text{CH}—\text{COOH} + \text{H}_2\text{S}$$
$$| \qquad\qquad\qquad\qquad\qquad\qquad\qquad |$$
$$\text{NH}_2 \qquad\qquad\qquad\qquad\qquad\qquad \text{NH}_2$$

Thus both inorganic sulfur and nitrogen can be formed, or only inorganic sulfur. Since serine sulfhydrase is reversible, it can also participate in the assimilation of sulfur. Methionine can be degraded with the formation of the mercaptan (CH_3SH) and NH_3. This is an example of the production of odoriferous volatile sulfur compounds in nature.

Plants and microorganisms generally utilize SO_4^{2-} to form sulfur-containing amino acids and other less-reduced substances. This involves assimilatory SO_4^{2-} reduction. A generalized scheme for SO_4^{2-} assimilation in nature is given in Fig. 13.5, showing cystine biosynthesis as found in *Escherichia coli*. Once SO_4^{2-} is taken across the cell membrane, the first reaction is the formation of APS by reaction with ATP. This reaction requires considerable energy and must be linked to the second reaction by APS kinase, which functions at low substrate concentrations. In a third mechanism, inorganic pyrophosphatase cleaves the pyrophosphate released in the first reaction. The overall reaction involves the use of two ATP molecules for every molecule of PAPS formed:

$$2 \text{ ATP} + SO_4^{2-} \rightarrow \text{ADP} + \text{PAPS} + PP_i$$

The PAPS can then be utilized to build organic sulfate esters or reduced and added to serine to produce cysteine. An alternate pathway for assimilatory SO_4^{2-} reduction, first found in *Chlorella* and spinach chloroplasts, involves the transfer of sulfur by bound intermediates, with APS rather than PAPS as a substrate for further reduction. This latter pathway has also been found to exist in *Escherichia coli*.

Mechanisms for the synthesis of methionine in plants involve transsulfuration and direct sulfuration leading to homocysteine, as an alternate pathway. General knowledge of the incorporation of sulfur into plants is lacking. When plants absorb SO_4^{2-} in excess of that required for incorporation into amino acids as precursors for protein synthesis, excess SO_4^{2-} (or in the case of Brassicaceae, glucosinolates) accumulates in the tissue. The concentration of plant HI-reducible sulfur is an excellent way for determining sulfur deficiency in soil. Studies have shown that when HI-reducible sulfur in wheat constitutes 10% or more of the total sulfur, the plants do not respond to additional sulfur. For rapeseed, the critical value is about 27%.

Figure 13.5. Activation of sulfate, ester sulfate formation, and the synthesis of cysteine by assimilatory sulfate reduction. APS, Adenosine -5'-phosphosulfate; PAPS, 3' Phosphoadenosine -5-phosphosulfate. (Adapted from Postgate, 1968, and Roy and Trudinger, 1970.)

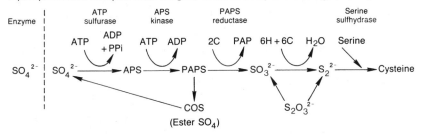

Oxidation of Inorganic Sulfur Compounds in Nature

Under aerobic conditions, reduced sulfur is oxidized through a variety of intermediates by both chemical and biological pathways. Oxidation states range from $+6$ in SO_4^{2-} to $+4$, $+2$, 0, and -2 in H_2S and its derivatives. However, because the intermediates can be chemically oxidized, only H_2S, S^0, and SO_4^{2-} accumulate during microbial growth. The nature of the oxidation pathway has not been clearly delineated, but it appears that S^0 is not an intermediate in oxidation of H_2S but is produced as a storage product. Oxidation of sulfur in soil generates acidity, and depending on the soil involved, this may be either helpful or harmful.

Oxidation of inorganic sulfur compounds in soils, waters, and sediments can be carried out by a diverse group of organisms. These can be subdivided into those growing at neutral pH and the forms that live at acidic pH values. The latter can also use ferrous iron as an electron donor, thus closely coupling sulfur and iron transformations. The following organisms are sulfur oxidizers.

Thiobacilli

Chemolithotrophic bacteria of the genus *Thiobacillus* are Gram-negative (G^-), nonsporulating rods. Thiobacilli oxidize H_2S, elemental S, thiosulfate, and polythionite. The following six species have been studied in most detail:

1. *Thiobacillus thiopurus*. It has a pH growth range of 5 to 8, with an optimum near neutrality.

2. *Thiobacillus denitrificans*. It is very similar to *T. thiopurus*. The organism can substitute NO_3^- for O_2 as an electron acceptor:

   $$2 NO_3^- + S + H_2O + CaCO_3 \rightarrow CaSO_4 + N_2$$

   The N_2 is lost as a gas to the atmosphere.

3. *Thiobacillus thioxidans*. It is usually characterized as the type species. It has a pH growth range of 2 to 5. It is a motile, strictly aerobic organism capable of carrying out the reactions shown in Fig. 13.6.

4. *Thiobacillus ferrooxidans*. It is a strictly aerobic organism with a pH growth range of 1.5 to 5. It also oxidizes ferrous iron (Fe^{2+}) as a source of energy and is of major significance in the production of acid mine water and in the commercial leaching of ores.

5. *Thiobacillus intermedius*. A facultative lithotroph with a pH growth range of 3 to 7 and capable of using $S_2O_3^{2-}$ as an electron donor. Growth is stimulated in the presence of organic matter.

Figure 13.6. Pathway of dissimilatory sulfate reduction in *Desulfovibrio*. (Adapted from Roy and Trudinger, 1970.)

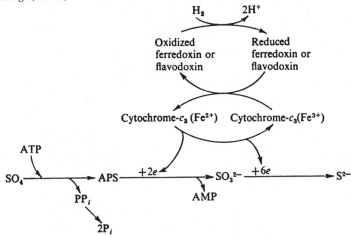

6. *Thiobacillus thiocyanoxidans*. It is very similar to *T. thiopurus*. For growth on sulfur or $S_2O_3^{2-}$, carbon and nitrogen must be supplied. When grown on thiocyanate, such as that found in sewage effluent, the following reaction occurs:

$$C\!\!-\!\!N\!\!-\!\!S^- + 2\,O_2 + 2\,H_2O \rightarrow SO_4^{2-}\ NH_4^+ + CO_2 + 220\ kcal$$

Sulfolobus

Sulfolobus comprises sulfur-oxidizing, thermophilic organisms that grow over a pH range of 2 to 5 and temperatures of 60 to 80°C. They therefore are of significance in sulfur-rich geothermal springs. These generally spherical organisms with distinct lobes, are capable of oxidizing H_2S. The cells also adhere tightly to sulfur crystals.

Thiomicrospora

Thiomicrospora comprises small spiral bacteria capable of passing through a 0.22-μm filter. One of the two species known can grow anaerobically, with NO_3^- as an electron acceptor.

Thermothrix

Thermothrix comprises filamentous bacteria growing in hot springs near a neutral pH and can utilize either O_2 or NO_3^- as an electron acceptor. This genus, as is also true for *Sulfolobus*, is capable of chemoorganotrophic growth.

Beggiatoa and *Thiothrix*

Members of the genera *Beggiatoa* and *Thiothrix* are colorless cells, growing in trichomes, showing a gliding motion, and, when grown in the presence of H_2S, containing sulfur granules. They occur in both freshwater and marine environments. Studies have indicated that although sulfur is metabolized, the organisms are primarily organotrophic. *Beggiatoa* has been found to oxidize HS^- in the rice rhizosphere, thus protecting the plant from this toxic compound.

Photolithotrophs

The green and purple bacteria represent a diverse morphological group, including cocci, vibrios, rods, spirals, and budding and gliding organisms. They have been discussed in Chapter 10. Photosynthetic sulfur oxidation is accomplished under anaerobic conditions by purple bacteria of the families Rhodosporillaceae and Chromatiaceae. The organisms range in color from bluish violet through purple, deep red, and orange-brown. The bacteria occur as single cells in the form of spheres, short rods, and spirals. They contain bacteriochlorophylls and carotenoids.

The green bacteria are small bacteria of varying morphology and belonging to the families Chlorobaceae and Chloroflexaceae. They have a green color due to the presence of chlorophyll but can also be brown if carotenoids are present. They frequently deposit elemental sulfur outside the cells. Both groups of photosynthetic sulfur bacteria are commonly found in mud and stagnant waters containing H_2S and exposed to light. They also occur in sulfur springs and saline lakes as a colored layer under salt deposits. There they reoxidize H_2S, coming from lower anaerobic layers. Their use of sunlight as a source of energy and H_2S as a source of electrons can be shown, as follows:

$$CO_2 + 2\ H_2S \xrightarrow{\text{light energy}} CH_2O + 2\ S + H_2O$$

$$3\ CO_2 + 2\ S + 5\ H_2O \xrightarrow{\text{light energy}} 3\ CH_2O + 2\ H_2SO_4$$

Heterotrophic Bacteria

Bacteria, such as *Arthrobacter, Bacillus, Micrococcus, Mycobacterium,* and *Pseudomonas,* some actinomycetes, and some of the fungi also oxidize inorganic sulfur compounds. Apparently no energy is made available to the organisms, and the transformations are incidental to the major metabolic pathways. These organisms are generally more numerous in soils than the chemolithotrophs and photolithotrophs. They may be the primary oxidizers of sulfur in neutral and alkaline soils until the pH is lowered sufficiently to permit oxidation by the chemotrophic Thiobacillaceae.

Since much of the soil organic sulfur in aerated soils is already in the sulfate ester form, only small concentrations of carbon-bonded sulfur are available for oxidation, and therefore, only low numbers of sulfur-oxidizing bacteria are normally present. One method of ameliorating sulfur deficiency in soil involves the addition of elemental sulfur. An adequate population of sulfur oxidizers is essential if the added sulfur is to be oxidized within a reasonable time. Field studies have indicated that lags in oxidation of sulfur added at planting time are often sufficiently long that the current crop fails to receive benefit.

Reduction of Inorganic Forms of Sulfur

The process known as respiratory SO_4^{2-} reduction is usually mediated by anaerobic, organotrophic organisms that use organic acids, alcohols, and often, H_2 as electron donors. These organisms are responsible for sulfide formation in waterlogged soils and sediments; they use SO_4^{2-} and other forms of inorganic sulfur as electron acceptors. Dissimilatory SO_4^{2-} reduction is now recognized in eight genera of bacteria. The best known of these is *Desulfovibrio,* a polarly flagellated G^- curved rod that occurs widely in soils and sediments. The genus *Desulfotomaculum* includes straight or curved G^- rods that produce endospores. Members of other genera have been isolated from marine or freshwater anaerobic environments. *Desulfomonas,* which consist of large, nonmotile, non-spore-forming rods, has physiological characteristics similar to *Desulfovibrio* and *Desulfomataculum*. Its activity in soil is unknown, as the intestinal tract has been its chief source of isolation.

Although the mechanisms of reduction are not fully resolved, the scheme shown in Fig. 13.7 suggests some of the possible steps. Electrons are transferred through reduced ferroxidan and flavodoxins to cytochromes, and SO_4^{2-} is activated by sulfate adenyltransferase and APS reductase, with the final product being the reduced sulfide. Most SO_4^{2-} reducers contain the electron carrier desulfovividin, which can be detected by its red fluorescence under alkaline conditions. The ability of many SO_4^{2-} reducers to utilize H_2 involves electron transport chains, which probably results in the generation of membrane potentials, as in respiratory schemes discussed in Chapter 2.

Sulfate-reducing bacteria are found over an extensive range of pH and salt concentrations, in saline lakes, evaporation beds, deep-sea sediments, and oil wells. The organisms can tolerate heavy metals and dissolved sulfide concentrations up to 2%. Although sulfate-reducing bacteria are largely organotrophic, in that most of the carbon fixed is derived from organic matter, some organic molecules, such as low molecular weight fatty acids (butyric, propionic, and acetic acids), are inhibitory to growth. In such cases, only certain organic substrates, such as lactic and pyruvic acids, are utilized as electron donors, as shown in the following:

$$2 \ CH_3CHOHCOOH \ + \ SO_4^{2-} \rightarrow 2 \ CH_3COOH + HS^- + H_2CO_3 + HCO_3^-$$
lactic acid

$$4 \ CH_3COCOOH \ + \ SO_4^{2-} \quad \rightarrow 3 \ CH_3COOH + CH_3COO^- + 4 \ CO_2 + HS^-$$
pyruvic acid

In some cases, in which H_2 can act as an electron source, CO_2 can be fixed, although small concentrations of complex organic compounds are required for growth. These compounds may act as chelating agents for Fe^{2+} to prevent its precipitation as FeS in the presence of high concentrations of HS^-.

When *Desulfovibrio* utilizes H_2, the following reactions occur:

$$S_2O_3^{2-} + 4 \ H_2 \rightarrow 2 \ HS^- + 3 \ H_2O$$

$$S_4O_6^{2-} + 9 \ H_2 \rightarrow 2 \ HS^- + 2 \ H_2S + 6 \ H_2O$$

Although most SO_4^{2-} reduction requires NH_3 as a nitrogen source, certain isolates reduce N_2 to NH_3. Therefore, they can grow in a CO_3^{2-}–SO_4^{2-} medium under atmospheres of N_2 and H_2.

The environmental consequences of SO_4^{2-} reduction are the following:

1. Reduction of SO_4^{2-} in sewage systems leads to the formation of sulfides. This has been implicated in the corrosion of stone and concrete. H_2S, with an odor of rotten eggs, clogs and corrodes pumps and sewage distribution systems. Its presence also stops the normal decomposition reactions required for sewage treatment. One solution for septic tanks is to add NO_3^- to the system. This acts as a more readily available electron acceptor, eliminating the production of H_2S.

2. Lagoons and stagnant lakes in which SO_4^{2-} is reduced to H_2S in the sediment can reoxidize the sulfur through the action of photosynthetic sulfur bacteria. Thiobacilli are not involved, as the system is anaerobic. As saline lakes dry out and the photolithotrophic sulfur oxidizers are no longer active, H_2S can escape to the environment.

3. Under anaerobic conditions, Fe^{3+} and SO_4^{2-} are reduced to Fe^{2+} and HS^-. These ions form a black precipitate of ferrous sulfide, which can further react to form a number of sulfide minerals. The reduction process increases the pH of the system. In China this principle is utilized to increase the pH of acidic soils. The reaction is

$$Na_2SO_4 + Fe(OH)_3 + 9 \ H \rightarrow FeS^+ + 2 \ NaOH + 5 \ H_2O$$

The production of *akiochi* disease in rice has been mentioned. This can be eliminated by the addition of NO_3^-, and to some extent, by the utilization of fertilizer compounds that do not contain sulfur.

4. Sulfur reduction in the geological past has led to the high concentrations of reduced sulfur in oil and coal fields. Unless removed during combustion, this sulfur leads to major pollution problems.

References

Brock, T. D., Smith, D. W., and Madigan, M. I. (1984). "Biology of Microorganisms." Prentice-Hall, Englewood Cliffs, New Jersey.

Postgate, J. R. (1968). Inorganic S chemistry. *In* "The S Cycle" (G. Nickless, ed.). Am. Elsevier, New York.

Roy, A. B., and Trudinger, P. A. (1970). "The Biochemistry of Inorganic Compounds of Sulfur." Cambridge Univ. Press, London and New York.

Saggar, S., Bettany, J. R., and Stewart, J. W. B. (1981). *Soil Biol. Biochem.* **13**, 493–505.

Winogradsky, S. (1887). Ueber Schwefelbacterien. *Bot. Ztg.* **45**, 489–610.

Winogradsky, S. (1889). Recherches physiologiques sur les sulphobacteries. *Ann. Inst. Pasteur Paris* **3**, 49–64.

Zehnder, A. J. B., and Zinder, S. H. (1980). The S cycle. *In* "Handbook of Environmental Chemistry" (D. Hutzinger, ed.), Vol. 1, pp. 103–145. Springer-Verlag, Berlin and New York.

Supplemental Reading

Adams, F., and Rawajfih, Z. (1977). Basaluminite and alunite: A possible cause of SO_4^{2-} retention by acid soil. *Soil Sci. Soc. Am. J.* **41**, 686–691.

Atlas, R. (1984). "Microbiology: Fundamentals and Application." Macmillan, New York.

Bettany, J. R., and Stewart, J. W. B. (1982). Sulfur cycling in soils. *Proc. Int. Sulfur Inst.* **2**, 767–785.

Chae, Y. M., and Lowe, L. E. (1980). Distribution of lipid S and total lipids in soils of British Columbia. *Can. J. Soil Sci.* **60**, 633–640.

Cooper, P. J. M. (1972). Aryl sulfatase activity in northern Nigerian soils. *Soil Biol. Biochem.* **4**, 333–337.

Ehrlich, H. L. (1981). "Geomicrobiology," pp. 251–280. Dekker, New York.

Fitzgerald, J. W. (1978). Naturally occurring organo-sulfur compounds in soils. *In* "Sulfur in the Environment. Part II. Ecological Impacts" (J. O. Nriagu, ed.), pp. 391–443. Wiley, New York.

Goldhaber, M. B., and Kaplan, I. R. (1975). The S cycle. *In* "The Sea" (E. D. Goldberg, ed.), Vol. 5, pp. 569–655. Wiley, New York.

Maynard, D. G., Stewart, J. W. B., and Bettany, J. R. (1984). Sulfur cycling in grassland and parkland soils. *Biogeochemistry* **1**, 97–111.

Neptune, A. M. L., Tabatabai, M. A., and Hanway, J. J. (1975). Sulfur fractions and carbon–nitrogen–phosphorus–sulfur relationships in some Brazilian and Iowa soils. *Soil Sci. Soc. Am. Proc.* **39**, 51–55.

Pepper, I. L., and Miller, R. H. (1978). Comparison of the oxidation of thiosulfate and elemental S by two heterotrophic bacteria and *Thiobacillus thiooxidans*. *Soil Sci.* **126,** 9–14.

Swank, W. T., and Fitzgerald, J. W. (1984). Microbial transformation of SO_4^{2-} in forest soils. *Science* **223,** 182–184.

Tabatabai, M. A., and Bremner, J. M. (1972). Forms of sulfur, and carbon, nitrogen and sulfur relationships in Iowa soils. *Soil Sci.* **114,** 380–386.

Chapter 14

Microbial Transformation of Metals

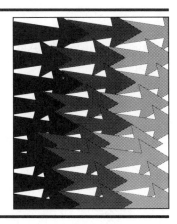

Introduction

Metal transformations can be divided into two major categories. The first involves oxidation or reduction of inorganic forms. Oxidation can serve as an energy source for microbial growth, whereas reduction involves the metal as an electron acceptor, providing some other energy source is available in an environment lacking other, more readily reduced substrates. The second category results in the conversion of the metals to organic forms, or the reverse conversion of organic to inorganic forms. These reactions are of significance in nature in a number of areas, including (1) soil formation, (2) reclamation of metals from low-grade ores, (3) acidification of mine waste water, (4) pollution of water supplies, and (5) production of metallic ores during geological time. Contamination of soils and ground waters with both organic and inorganic forms of metals, such as arsenic, mercury, selenium, and cadmium, causes problems with toxic-waste-disposal sites. Contamination also can occur near naturally occurring deposits where water movement results in surface and ground-water contamination.

The association of bacteria with deposits of bog iron was first described in 1838 (reviewed by Ehrlich, 1981). Bacteria were thought to contribute to the formation of the deposits. Winogradsky, in his early work (1888, 1949), believed that the bacterium *Leptothrix,* which forms a precipitate of ferric hydroxide, gains energy from the process of oxidizing ferrous (Fe^{2+}) to ferric (Fe^{3+}) iron. Microbial activity in manganese oxidation, which resembles that of iron, and precipitation was first described by another early soil microbiologist, M. W. Beijerinck, in 1903. Investigations of the origin of acid coal-mine drainage in the 1950s resulted in the discovery of the iron-oxidizing thiobacilli. It was demonstrated that these microorganisms could oxidize insoluble metal sulfides of iron and copper. This process leads to acidification and to solubilization of the metal (Weinberg, 1977).

Transformation of Iron

Iron can exist in oxidation states Fe^0, Fe^{2+}, and Fe^{3+}. Under acidic conditions, metallic iron (Fe^0) readily oxidizes to the ferrous state (Fe^{2+}). At pH values less than 5, the Fe^{2+} state is stable, but at pH values greater than 5 it is chemically oxidized to Fe^{3+}. In turn, Fe^{3+} is readily reduced under acidic conditions but is precipitated in alkaline solution. Since the oxidation of Fe^{2+} is spontaneous at pH values above 5.0, it is difficult to prove that the reaction is enzyme controlled at higher pH values. A Gram-negative motile rod, now designated *Thiobacillus ferroxidans*, mediates this reaction at acidic pH values and derives both energy and reducing power from the reaction. It can, therefore, be said to be a chemolithotroph, as defined in Chapter 4. Organisms previously described as *Ferrobacillus ferroxidans* and *F. sulfoxidans* are now considered to be *T. ferroxidans*.

Before Fe^{2+} is microbiologically oxidized, it is chelated by organic compounds to increase its availability. During oxidation, the electrons are moved through the electron transport chain, as described in Chapter 2, with cytochrome *c* being the point of entrance into the transport chain. Passage of protons through the cell membrane also is thought to occur and to provide further energy. The search for bacteria capable of oxidizing metals and releasing them from sulfide ores in biotechnological mining is leading to the discovery of previously undescribed organisms. Some of these are similar to *Sulfolobus*, as described in Chapter 13. Others will require new designations.

Precipitates of iron around filamentous, sheath-forming bacteria, such as the Chlamydobacteriaceae, of which *Leptothrix* (Fig. 14.1) is an example, can occur in well waters or in iron pipes. Another organism that causes precipitation of iron is *Crenothrix*. This is a nearly spherical, non-motile cell 2 μm in diameter. On division, it forms a chain of elongated cells. A mucilaginous, tubular sheath approximately 0.2 μm in thickness is exuded. The sheath, on hardening, becomes impregnated with $Fe(OH)_3$. *Crenothrix* grows in water containing low concentrations of Fe^{2+} and a small amount of dissolved O_2. Under these conditions it accumulates $Fe(OH)_3$. It also produces large gelatinous masses that can clog pipes. When the organisms die the decomposition of the organic substrate often results in a disagreeable odor, and the $Fe(OH_3)$ sheaths result in a red

Figure 14.1. Representative iron-precipitating and -oxidizing bacteria.

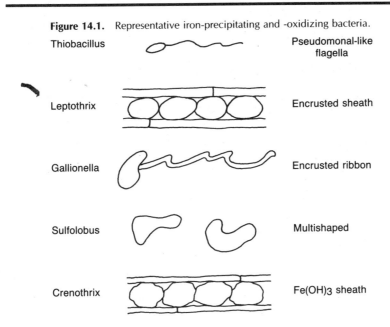

Thiobacillus — Pseudomonal-like flagella

Leptothrix — Encrusted sheath

Gallionella — Encrusted ribbon

Sulfolobus — Multishaped

Crenothrix — Fe(OH)$_3$ sheath

precipitate. Although there is no known specific cure, the iron-precipitating bacteria can be killed by treatment with agents such as Clorox.

The stalk-forming *Gallionella,* comprising bean-shaped, 0.5×2 µm cells growing in flat 200- to 300-µm mucilaginous ribbons encrusted with Fe(OH)$_3$, also are involved in iron precipitation. In addition to causing contamination in well waters and pipes, these organisms also cause odor and slime problems in Fe^{2+}-polluted streams and rivers.

In nature, the oxidation–reduction process over geological time proceeds through sediments to sedimentary rocks and results in the formation of minerals such as geothite and pyrite. In the presence of sulfate and sulfate-reducing bacteria, the reduced iron reacts with H$_2$S to form the mineral hydrotrilite, which in the geochemical cycle eventually forms pyrite (FeS$_2$). Sulfate is in excess in oceanic waters; therefore, this process maintains the iron content of the ocean water at very low concentrations.

The Fe^{3+} form, common in most soils, is insoluble. Uptake involves proton exudation (acidification) or reaction with organic compounds with strong affinity for Fe^{3+}. These organic compounds, known as ferrichromes or siderophores, often consist of hydroxamates, which are composed of a hydroxyl group closely associated with a nitrogenous portion of the molecule (Fig. 14.2A). This results a very strong affinity for the multivalent iron through the process of chelation. Soil solution concentrations of iron

Figure 14.2. Structure of the siderophores schizokinen (A) and mycobactin (B). Iron chelation sites are shown by asterisks. (From Byers and Arceneaux, 1977.)

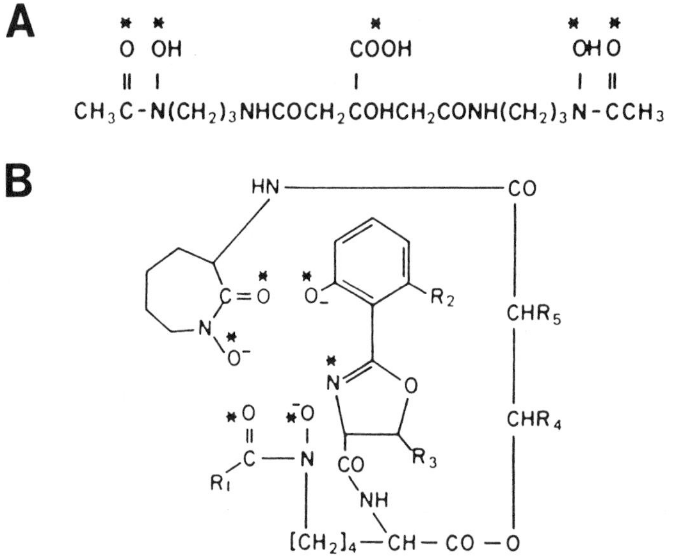

A

```
     *  *                    *                  *   *
     O  OH                   COOH               OH O
     ‖  |                    |                  |  ‖
CH₃C-N(CH₂)₃NHCOCH₂COHCH₂CONH(CH₂)₃N-CCH₃
```

B

less than optimal for plant growth occur in alkaline soil. The control of available iron concentrations in the rhizosphere by siderophore-producing bacteria is therefore an important soil process since iron, a constituent of many enzymes, is required by all organisms. It has been postulated that siderophore-producing pseudomonads in the rhizosphere act in the control of root pathogens. Siderophores produced by certain pseudomonads are thought to lower the available iron concentrations to levels at which plant pathogens are inhibited but plant uptake is not.

Just as iron can be oxidized by the direct involvement of enzymes, or nonenzymatically by microorganisms that raise the redox potential or the soil pH, iron reductions can occur via a number of processes. Ferric iron can serve as a respiratory electron acceptor. It can also be reduced by reaction with microbial end products such as formate or H_2S. Fungi, such as *Alternaria* and *Fusarium,* and bacteria, such as *Bacillus,* have been found capable of Fe^{3+} reduction. The enzyme nitrate reductase may be involved in these reactions, and the presence of NO_3^- usually provides a high enough redox potential P_ε to inhibit iron reduction.

Iron oxidations and reductions play a major role in the soil formation process known as gleization. Under waterlogged conditions, a light gray, greenish color attributable to the presence of reduced iron is produced. The Fe^{3+} is reduced to Fe^{2+}, as an alternate electron acceptor to O_2. The gleyed soils usually have mottled spots of Fe_2O_3, where root channels or cracking have caused localized oxidation. The gleying reaction results in the sealing of soils under ponds. Soviet workers have found that if a layer of straw of approximately 15 cm is buried in the bottom of a dugout and covered with another 15 cm of soil, the decay of the organic matter results in the reduction of Fe^{3+} to Fe^{2+} and of SO_4^{2-} to S_2^{2-}. FeS_2 is precipitated and the soil colloids are peptized. This results in the sealing of all cracks and is a much cheaper process than the use of plastic sheets. It appears permanent as long as the soil continues to be waterlogged and can be used to seal man-made lakes and, possibly, waste-disposal sites.

Translocation of iron is involved in the soil podzolization process in which iron and organic-matter-rich subsurface horizons are formed. In this process, certain microorganisms produce organic acids and chelating agents that solubilize iron. The soluble iron–organic complex is leached through the soil profile by percolating water during rainfall. Precipitation of this iron in the soil B horizon results in the production of podzolic soils (spodosols). These consist of a leached upper layer and an organic matter iron-rich lower (B) horizon. The soil-forming process in which the iron is primarily left behind, but silica is translocated, results in the laterization process often found in tropical soils.

Anaerobic Corrosion of Iron Pipes

Microorganisms participate in corrosion processes, as shown in Table 14.1. In anaerobic corrosion, the metal surface, acting as an anode in an electrochemical reaction, is transformed to Fe^{2+}. At the cathode site, an equivalent number of H^+ ions are produced. The anaerobic, sulfate-reducing bacterium *Desulfovibrio* produces S^{2-}. This reacts with Fe^{2+} to produce FeS. At the same time, hydroxyls from water react with the H^+ ions. The overall reaction is

$$4\ Fe + SO_4^{2-} + 4\ H_2O \rightarrow FeS + 3\ FeOH^{2-} + 5\ OH^-$$

The conditions required for the above reaction include anaerobic sites at redox potentials less than 400 mV ($P_\varepsilon = 6.8$), a pH greater than 5.5, a low free O_2 content, and the presence of SO_4^{2-}. Under these conditions an iron pipe of 3-mm wall thickness can be corroded in 5 to 7 years. This is one of the most expensive microbial reactions in nature, for it means that buried iron pipes must be continually replaced unless they are protected. Protection includes careful wrapping with asphalt and plastic materials and by maintaining a small electrical current along the pipe (cathodic protection) so that the corrosion cannot be initiated by the formation of the electrode half-cell.

Table 14.1
Some Bacteria Involved in Corrosion Processes[a]

Organism	Oxygen requirement	Inorganic components	Metabolic end products	Habitat	Optimal range	
					Temperature	pH
Sulfate reducing						
Desulfovibrio desulfuricans	Anaerobic	Sulfate, thiosulfate	Hydrogen sulfide	Water, soil, mud, oil reservoir	25–30	6–7.5
Sulfur oxidizing						
Thiobacillus thiooxidans	Aerobic	Sulfur, thiosulfate	Sulfuric acid	Soil, water	28–30	2–4
Thiosulfate oxidizing						
Thiobacillus thioparus	Aerobic	Thiosulfate, sulfur	Sulfur, sulfuric acid	Soil, water, mud, sewage	28–30	7
Iron bacteria						
Crenothrix, Leptothrix, Gallionella	Aerobic	Iron, manganese	Ferric or manganese oxides	Water	25	8
Nitrate reducing						
Thiobacillus dentrificans	Facultative	Thiosulfate, sulfur, sulfide	Sulfate	Soil, mud, peat, water	30	7–9
Hydrogen utilizing						
Hydrogenomonas	Microaerophilic	Hydrogen	Water	Soil, water	28–30	7

[a]From Atlas (1984).

Other corrosion processes, as shown in Table 14.1, include those by aerobic and facultative bacteria, such as *Gallionella* and *Leptothrix*. The corrosion by these organisms, although very noticeable, is not as economically important as the anaerobic processes. Indirect reactions include those causing acidic conditions attributable to sulfur oxidations by thiobacilli.

Transformation of Manganese

Bacterial oxidation of Mn^{2+} occurs in both soil and sediments. On ocean bottoms, microbial activity has been implicated in the formation of ferromanganese nodules. The chemical oxidation of Mn^{2+} occurs only above pH 8. Oxidation of manganese in the environment at neutral or acidic concentrations must, therefore, be microbiologically mediated. A number of soil bacteria and fungi can oxidize manganese ions. One specific example is *Arthrobacter*, which appears to do this by constitutive enzymes. The sheathed, iron-oxidizing bacterium *Leptothrix* (formerly known as *Sphaerotilus*) also oxidizes Mn^{2+} to Mn^{4+} and accumulates MnO_2 in its sheaths and filaments.

Manganese-oxidizing bacteria can be responsible for manganese deficiency symptoms in plants. Susceptible varieties of oats develop grayspeck disease; this can be overcome by the addition of manganese. Under certain soil conditions the rhizosphere bacteria oxidize Mn^{2+} to Mn^{4+} and deposit MnO_2 on the outside of the root. Examination of the roots under such conditions shows a black precipitate of MnO_2. The easiest method of control is by genetic selection of the host plant, which in turn controls the rhizosphere population. Alternatively, manganese toxicity from excess manganese can occur in acidic soils. The precipitation of manganese in the filaments of mycorrhizal fungi before it enters the plant has been found to allow the growth of manganese-sensitive plants under such conditions.

Transformation of Mercury

Mercury is the metal most commonly involved in health problems involving metals. Large amounts of metallic mercury (Hg^0) were deposited in certain streams by commercial chloralkali plants that utilized metallic mercury in the electrolysis of NaCl in the production of NaOH and bleach (Cl_2 and OCl). This resulted in the contamination of river sediments. However, the mercury did not stay in the metallic form. Microorganisms in the sediment transformed the metallic mercury to methylmercury (CH_3Hg^+) and dimethylmercury (CH_3HgCH_3). The methylated forms are not only volatile, they are also more readily absorbed and retained in animal tissue than the metallic form. This has led to incidences of human mercury poisoning; unfortunately, a number of areas of the globe must still restrict the consumption of fish because of their high mercury concentrations.

Metallic mercury has a high vapor pressure, and approximately 10^4 metric tons of mercury is found in the atmosphere. This comes from mercury-containing minerals and the combustion of fossil fuels. Coal contains 1200–21,000 ppb mercury, and the burning of crude oil and coal releases between 2×10^4 and 7×10^4 metric tons of mercury per year to the global system. Human activities since the late 1800s have released 10^6 metric tons of mercury. This eventually finds its way to the oceans. Less than 1% of the 10^8 metric tons of mercury found in the oceans, however, is of anthropogenic origin. It is very difficult to control the cycling of mercury; however, as we gain a greater understanding of the microbiological reactions involved, we shall know more about its presence in soils and in plant and animal products and can act in a responsible manner to reduce its pollution effects.

References

Atlas, R. M. (1984). "Microbiology: Fundamentals and Applications." Macmillan, New York.

Byers, B. R., and Arceneaux, S. L. (1977). Microbial transport and utilization of iron. In "Microorganisms and Minerals" (R. P. Weinberg, ed.), pp. 215–249. Dekker, New York.

Ehrlich, H. L. (1981). "Geomicrobiology." Dekker, New York.

Waksman, S. A. (1946). Sergei Nikolaevitch Winogradsky: The story of a great bacteriologist. Soil Sci. 62, 197–226.

Weinberg, R. P. (1977). "Microorganisms and Minerals." Dekker, New York.

Winogradsky, S. N. (1949). "Microbiologie du sol; Problèmes et méthodes." Masson, Paris.

Supplemental Reading

Iverson, W. P. (1974). Microbial corrosion of iron. In "Microbial Iron Metabolism: A Comprehensive Treatis" (J. B. Neilands, ed.), pp. 476–517. Academic Press, New York.

Konetzka, W. A. (1977). Microbiology of metal transformations. In "Microorganisms and Minerals" (E. Weinberg, ed.), pp. 317–341. Dekker, New York.

Sajic, J. E. (1969). "Microbial Biogeochemistry." Academic Press, New York.

Starkey, R. L., and Halvorson, H. O. (1927). Soil Sci. 24, 381–402.

Summers, N. O., and Silver, S. (1978). Microbial transformations of metals. Annu. Rev. Microbiol. 32, 638–672.

Glossary

Concepts from a number of scientific disciplines are discussed in this book. The glossary is provided with the hope that the terminology associated with diverse disciplines will not be an impediment to the reader. The definitions are written to present the essential aspects of a given term in the context in which it is used. More detailed definitions may be obtained from the appropriate supplementary reading.

Akinete, a nonmotile spore formed singly within a cell and with the spore wall fused with the parent cell wall.

Ammonification, the biochemical process whereby ammoniacal nitrogen is released from nitrogen-containing organic material.

Antibiotic, a substance produced by one organism that is inhibitory or biocidal to another organism.

Assimilation, the incorporation of inorganic or organic substances into cell constituents.

Assimilatory reduction of nitrate, the conversion of nitrate to reduced forms of nitrogen for use in the synthesis of amino acids and proteins.

APS, adenosine 5-phosphosulfate; the activated sulfate involved in reduction of sulfates.

ATP, adenosine triphosphate; the major carrier of phosphate and growth energy, composed of adenine and three phosphate groups.

Autochthonous, growing slowly in soil containing no easily oxidizable substrate; see *Zymogenous*.

Autotrophic, obtaining growth energy from inorganic sources and cell carbon from CO_2.

Bacteriophage, a virus that infects bacteria.

Bacteroid, irregularly shaped cell form that certain bacteria can assume under some conditions, e.g., *Rhizobium* and *Bradyrhizobium* in root nodules.

Benthic, aquatic organisms that grow on or inhabit the bottoms of lakes and oceans.

Biological denitrification, see *Denitrification*.

Biomass, the total mass (dry weight) of living organisms.

Chemodenitrification, the nonbiological process leading to the production of gaseous nitrogen from oxidized nitrogen, principally through nitrite instability.

Chemotrophic, obtaining growth energy from chemical materials.

Chlamydospore, a thick-walled multinucleate asexual spore developed from hyphae but not on basidia or conidiophores; produced by many parasitic and mycorrhizal fungi.

Coenocyte, a multinucleate cell or protoplast in which nuclear divisions have not been followed by cytoplasmic cleavage.

Conidium, a thin-walled secondary asexual spore borne terminally on a specialized hypha termed a conidiophore.

Cytokinin, substituted adenines that are growth regulators in plants.

Deamination, the removal of an amino group ($-NH_2$), freeing ammonia.

Denitrification, the biological reduction of nitrate or nitrite to molecular nitrogen or to oxides of nitrogen; this process is also termed enzymatic denitrification and biological denitrification.

Diazotroph, an organism or an association of organisms that can utilize dinitrogen (N_2) for growth.

Dissimilatory reduction of nitrate, the use of nitrate by organisms as an alternate electron acceptor in the absence of O_2, and causing reduction of nitrate to ammonium.

DNA, deoxyribonucleic acid, a type of cell nucleic acid carrying genetic information and containing adenine, guanine, cytosine, thymine, and 2-deoxyribose, linked by phosphodiester bonds.

Ectomycorrhiza, a fungus–root association in which the fungal hyphae form a mantle on the roots; mycelial strands penetrate inward between cortical cells and outward into the surrounding soil.

Endomycorrhiza, a fungus–root association with extensive intercellular and intracellular penetration of the root by the fungus; mycelial strands also extend outward into the surrounding soil.

Endophyte, an organism growing within a plant. The association may be parasitic or symbiotic.

Enzymatic denitrification, see *Denitrification.*

Eukaryote, an organism with a membrane-bound nucleus within which the genomes are stored as chromosomes, e.g., fungi, protozoa, plants.

Fermentation, energy-yielding metabolism involving oxidation–reduction reactions in which an organic substrate or the organic compounds derived from the substrate serve as the primary electron donor and the terminal electron acceptor.

First-order reaction, one in which the rate of transformation of the substrate is proportional to the substrate concentration.

Fulvic acid, the mixture of organic substances extractable from soil or sediment by weak alkali and not precipitated on acidification of the extract.

Gamete, a sexually reproductive cell or unisexual protoplasmic body incapable of giving rise to another individual until after conjugation with another gamete and the joint production of a zygote.

Gametangium, a differentiated structure in which gametes are produced, or the contents of which may function as a gamete.

Gene splicing, the cutting of a gene out of the DNA sequence of one organism and splicing it into the DNA of another.

Genetic engineering, manipulating the genetic structure of an organism by transfer of genes from one organism to another; see *Gene splicing, Transduction, Transformation.*

Gleization, a soil-forming process under poor drainage conditions, resulting in the reduction of iron and other elements and causing gray coloration and soil mottling.

Hartig net, the intercellular net of hyphae between the cortical cells of a root in an ectomycorrhiza.

Heterocyst, a large, thick-walled cell occurring every 10 to 15 cells in filaments of cyanobacteria; such cells separate intervening hormogonia and often are involved in nitrogen fixation.

Heterotrophic, obtaining growth energy and most cell carbon from organic materials; see *Wood–Werkman phenomenon.*

Hormogonium, a segment of cells, marked off by heterocysts in a cyanobacterial filament, capable of detachment and development into a new filament.

Humic acid, the dark-colored fraction of the soil humus extractable by weak alkali and precipitating following acidification of the extract.

Humin, the fraction of the soil humus that is not dispersable by weak alkali or pyrophosphate.

Humus, the dark-colored major fraction of the soil organic matter, formed during the decomposition of organic residues and containing humic and fulvic acids and other poorly defined or unknown substances relatively resistant to decomposition.

Hyphae, branched or unbranched filaments that form the vegetative form, as in filamentous fungi, algae, and bacteria.

Hyperbolic reaction, one having a rate constant for which the plot of the product formed versus time yields a curve approaching some maximum value, best described by a hyperbolic equation; see *Langmuir equation, Michaelis–Menten equation, Monod equation.*

Immobilization, the conversion of an element from the inorganic to the organic form by microorganisms or plants.

Karyogamy, the union of gamete nuclei to form a zygote nucleus.

km, the Michaelis constant, the substrate concentration of half the maximum velocity of an enzyme reaction.

Langmuir equation, the hyperbolic equation commonly used in physical chemistry to describe adsorption phenomena; see *Hyperbolic reaction.*

Laterization, a soil-forming process in the humid tropics leading to soils with a low silica/sesquioxide ratio in the clay fractions, low clay activity, low content of most primary minerals and soluble constituents, a high degree of aggregate stability, and usually red in color.

Leghemoglobin, an iron–heme protein controlling the oxygen content in legume nodules.

Lithotrophic, organisms obtaining their cell carbon from CO_2, and not from organic matter.

Lysozyme, enzyme that hydrolyses peptidoglycans and thus degrades bacterial cell walls.

Mesofauna, the soil-dwelling metazoans large enough to cause disturbance of the soil pores when moving through soil; includes nematodes, mites, collembola, enchytraeids, ants, insect larvae, and numerous others.

Microbiota, the soil biota having microscopic dimensions; includes the soil microflora, protozoa, and microscopic metazoa.

Michaelis–Menten equation, see *Hyperbolic reaction.*

Microfauna, the soil protozoa and other fauna of microscopic dimensions and generally capable of moving through existing soil pores.

Mineralization, the conversion of an element from an organic to an inorganic form; gross mineralization is the total amount converted, and net mineralization, the gross minus the growth demand or immobilization by the decomposer organisms, usually measured in laboratory incubations.

Mollisol, in the U.S. Department of Agriculture soil classification system, a soil that has a dark-colored high organic matter, highly base-saturated surface horizon, usually associated with grasslands.

Monod equation, the hyperbolic equation commonly used to describe microbial growth rate; see *Hyperbolic reaction.*

Mycelium, the collective term for fungus filaments or hyphae; the assimilative phase of a body of hyphae, as opposed to reproductive organs or phases.

Myxamoeba, the naked cell formed when a zoospore sheds its flagella and becomes ameboid.

NADP, the phosphorylated form of nicotine adenine dinucleotide (NAD), formed when the coenzyme serves as an electron donor in oxidation–reduction reactions.

ng, nanogram, 10^{-9} grams.

Nitrate respiration, see *Dissimilatory reduction of nitrate.*

Nitrification, the biological oxidation of ammonium and nitrite to nitrate.

Nitrogen mineralization, see *Mineralization, Ammonification.*

Nucleic acid, a large macromolecule composed of purine and pyrimidine bases, sugars, and phosphoric acid.

Organotrophic, obtaining growth energy and cell carbon from organic materials.

PAPS, 3-phosphoadenosine 5-phosphosulfate.

Peptidoglycan, the rigid acetylglucosamine, acetylmuramic acid–amino acid layer of bacterial cell walls.

Permease, an enzyme that increases the rate of transfer of a substance across a membrane.

Phototrophic, obtaining growth energy from light by means of photosynthetic processes.

Phyllosphere, the collective surfaces of the aboveground parts of live plants.

P_i, inorganic phosphate.

Plankton, the aquatic organisms that are free-floating in bodies of water.

Plasmodium, a body of naked, multinucleated protoplasm exhibiting amoeboid motion.

Podzolization, a soil-forming process occurring in temperate to temperate humid climates under coniferous or mixed coniferous and deciduous forest and characterized by a highly leached, whitish gray A_2 horizon.

PP_i, inorganic polyphosphate.

Prokaryote, an organism whose genomes are not contained in a nucleus, e.g., bacteria.

Respiratory reduction of nitrate, see *Dissimilatory reduction of nitrate.*

Rhizoplane, the collective surfaces of the roots of live plants.

Rhizosphere, the soil region immediately surrounding plant roots and in which microorganisms are affected by live roots.

RNA, ribonucleic acid, a linear polymer in which ribose residues are linked by 3'5'-phosphodiester linkages to nitrogenous bases such as adenine, guanine, uracil, or cytosine.

Septate, separated by cross walls.

Siderophore, a ferrichrome, a complex nitrogen-containing hydroxamate produced by organisms and having a strong chelating affinity for iron.

Soil organic matter, the organic fraction of soil including plant and animal residues, soil organisms, microbial metabolites, and humified constituents; usually determined on soils from which organic materials larger than 2 mm have been removed.

Spodosol, an order in the U.S. Department of Agriculture soil taxonomy in which a mineral illuvial soil horizon accumulates amorphous aluminum and organic carbon, and may or may not accumulate iron; occurs under leaching conditions in forests.

Sporocarp, a multicelled body capable of producing spores.

Substrate, a compound on which an enzyme (or organism) acts.

Symbiosis, the living together in intimate association of two dissimilar organisms, the collaboration being mutually beneficial.

Tg, terragram, 10^{12} grams or one million metric tons.

Thallus, the assimilative part of algae, lichens, fungi.

Transduction, the transfer of genes through virus vectors (bacteriophages) wherein virus DNA picks up host DNA and transfers it to another host organism.

Transformation, the transfer of DNA from one bacterium to another.

Wood–Werkman phenomenon, the uptake and incorporation into organic substances of CO_2 by heterotrophs; tracer data show that heterotrophs, although obtaining most of their cell carbon from organic matter, do obtain some from CO_2.

Zero-order reaction, one in which the rate of transformation of a substrate is independent of changes in substrate concentration.

Zoospore, a motile or swarm spore.

Zygospore, a thick-walled resting spore formed by conjugation of gametes or, in the Zygomycetes, by fusion of similar gametangia.

Zymogenous, subsisting on easily decomposable organics; responding quickly to freshly added organic materials.

Index